H. *Laurens van der Laan*

The Trans-Oceanic Marketing Channel
A New Tool for Understanding Tropical Africa's Export Agriculture

Pre-publication
REVIEWS,
COMMENTARIES,
EVALUATIONS . . .

"The literature on African export agriculture takes the world market as the starting point. But African farmers participate in the world market only indirectly, via a set of institutions and actors that form marketing channels. This book analyzes the channels through which African exports flow, following the product from origin to destination, and studies the objectives and operations of the actors who successively possess the product. Though the world market is permanent, the marketing channel is vulnerable and its continuity is never to be taken for granted.

Van der Laan uses marketing channel analysis to make sense of a century of change in the marketing of Africa's exports. He shows how technological innovation led to changes in commercial organization and practices, and examines the effects of different forms of marketing institutions for African farmers and traders. He identifies tensions between actors and the motivations for policy changes, both in the colonial and post-colonial periods. This book will be useful for anyone concerned with marketing institutions today. It raises issues that must be considered if privatization is not to lead to the collapse of agricultural exports. It will be of interest to economists, geographers, historians, and public policy analysts alike."

Barbara Grosh, PhD
Assistant Professor
of Public Administration,
Syracuse University,
Syracuse, NY

"**A**pproaches to development through export-led growth will never be quite the same again after the publication of this fine and exciting book. Laurens van der Laan writes from long experience of Africa, rejecting modish pessimism as to the continent's growth potential, and his analysis is highly relevant to many other Third World countries. Whereas past analyses have focused on country-to-country trade relations, van der Laan follows agricultural commodities along the marketing channels that take them from planting to consumption. With insights gleaned from New Institutional Economics, and with historical depth going back to the 1860s, the author makes some fascinating micro-economic analyses of what is normally considered a macro-economic phenomenon. He warns against a simplistic faith in the power of markets, and shows how these are structured by the complex institutions of commodity channels, both formal and informal. Particular emphasis is placed on transaction costs and the risks inherent in holding stocks. The preference of exporters for highly marketable commodities and rapid export velocity is explained in these terms, as are the advantages of commodity exchanges, and futures markets in importing countries are set out. Van der Laan writes with the burning conviction that knowledge of these factors can help to 'emancipate' Third World producers, middlemen, and officials, making them able to negotiate more effectively. While state action can be beneficial in spreading information and nurturing institutions, export marketing boards and international commodity agreements have proved to be broken reeds. Countries with a small 'fund of exporting competence' also face dire consequences if they persecute skilled commercial minorities, such as the South Asians and the Syro-Lebanese. This is a book that should be read with care by researchers and decision makers in Third World bureaucracies, donor governments, multilateral financial agencies, private aid organizations, Fair Trade groups, trading corporations, and faculties of economics and economic history."

William G. Clarence-Smith, PhD
Reader in the Economic History of Asia and Africa, School of Oriental and African Studies, University of London

More pre-publication
REVIEWS, COMMENTARIES, EVALUATIONS . . .

"This book provides a wide-ranging overview of the developments in Africa's agricultural commodity export marketing over the past century, not only for the main export crops, but also for often overlooked minor crops such as ginger and bananas. Changes in structures, actors, and policies are described against the background of the objective factors that drive behavior in an imperfect market, providing logical explanations for what otherwise would seem irrational marketing practices. One of the strengths of the book is the extensive discussion of changes in shipping, telecommunications, grading requirements, etc., and their impact on commodity trade.

The insights into what motivates marketing behavior are at times provocative, and useful not only for researchers, but also for policymakers and those who advise them. The key importance of access to credit and exposure to risks, as well as the positive role that traders play once they have built up marketing skills, are shown to be consistent features of trade in the period analyzed–even though now they are often overlooked. Commodity marketing policies in Africa are in a process of transformation; this book can help in ensuring that this transformation will be guided by economic reality, rather than, as happened too often in the past, ideology."

Lamon Rutten
Officer-in-Charge,
Commodity Marketing
and Risk Management,
United Nations Conference
on Trade and Development,
Geneva, Switzerland

"As a lawyer and legislative advisor to a state senator and state representative who concentrate their legislative efforts on health care for the elderly, I found this book an indispensable tool to analyze and contrast issues for proposed legislation."

Francisco Hernandez, Jr.
Attorney, Osuna & Hernandez,
Fort Worth, TX

The International Business Press
An Imprint of The Haworth Press, Inc.

The Trans-Oceanic Marketing Channel
A New Tool for Understanding Tropical Africa's Export Agriculture

THE INTERNATIONAL BUSINESS PRESS
Erdener Kaynak, PhD
Executive Editor

New, Recent, and Forthcoming Titles:

The Trans-Oceanic Marketing Channel
A New Tool for Understanding Tropical Africa's Export Agriculture

H. Laurens van der Laan

The International Business Press
An Imprint of The Haworth Press, Inc.
New York • London

Published by

The International Business Press, an imprint of The Haworth Press, Inc., 10 Alice Street, Binghamton, NY 13904-1580

Cover design by Monica Seifert.

Library of Congress Cataloging-in-Publication Data

Laan, H. L. van der.
    The trans-oceanic marketing channel : a new tool for understanding tropical Africa's export agriculture / H. Laurens van der Laan.
        p.   cm.
    Includes bibliographical references and index.
    ISBN 0-7890-0116-0
    1. Farm products–Africa–Marketing. 2. Export marketing–Africa. 3. Marketing channels–Africa. 4. Shipping–Africa. 5. Farm produce–Tropics–Marketing. 6. Export marketing–Tropics. 7. Marketing channels–Tropics. 8. Shipping–Tropics. 9. Tropical crops–Marketing. I. Title.
HD9017.A2L3 1997
382'.41'096–dc21   97-5150
                                                              CIP

CONTENTS

Tables, Figures, and Maps

Chapter 7

Chapter 9

Chapter 10

Chapter 12

Chapter 14

Chapter 15

Abbreviations

ACP	African Caribbean Pacific
AETC	African and Eastern Trade Corporation
APROMA	Association des Produits à Marché
CFDT	Compagnie française pour le développement des fibres textiles
CIF	cost, insurance, and freight
CFAO	Compagnie Française de l'Afrique Occidentale
CMC	conventional marketing channel
COD	cash on delivery
CPA	Cocoa Producers Alliance
CTC	colonial trading companies
EMB	Export Marketing Board
FAO	Food and Agriculture Organization of the United Nations
FAQ	fair average quality
FOB	free on board
HPS	Hand Picked and Selected
ICO	International Coffee Organization
IMC	international marketing channel
IMP	intermediate marketing point
KNCU	Kilimanjaro Native Co-operative Union
KPCU	Kenya Planters' Co-operative Union
KTDA	Kenya Tea Development Authority
LBA	Licensed Buying Agent
LDC	less developed country
MCA	Marketing Channel Analysis
NAHV	Nieuwe Afrikaansche Handels-Vennootschap
PMP	primary marketing point
SBS	sample before shipment
SCOA	Société Commerciale de l'Ouest Africain
SCP	structure, conduct, performance

SPS	Sudan Plantations Syndicate
TNC	trans-national corporation
TOMC	trans-oceanic marketing channel
UAC	United Africa Company
UFC	United Fruit Company
UNCTAD	United Nations Commission on Trade and Development
VMS	vertical marketing system
VOSMS	vertical ocean-straddling marketing system

MAP 1. African States, 1996

18 Djibouti	23 Mozambique
19 Uganda	24 Zimbabwe
20 Rwanda	25 Botswana
21 Burundi	26 Swaziland
22 Malawi	27 Lesotho

1 Western Sahara
2 Senegal
3 Gambia
4 Guinea-Bissau
5 Guinea
6 Sierra Leone
7 Liberia
8 Côte d'Ivoire
9 Burkina Faso
10 Ghana
11 Togo
12 Benin
13 Cameroon
14 Central African Republic
15 Equatorial Guinea
16 Gabon
17 Congo

xiii

ABOUT THE AUTHOR

H. Laurens van der Laan, PhD, is Senior Researcher at the African Studies Centre in Leiden, The Netherlands. He also teaches part-time for the Department of Marketing at the Agricultural University in Wageningen. Since 1975, Dr. van der Laan has performed comparative research on the marketing of African export crops, with marketing boards as his special focus. In 1983, he was the principal organizer of an international conference on "Marketing Boards in Tropical Africa," which led to his co-editing a book of that same name. He began his academic career in Sierra Leone, where he taught economics at the University of Sierra Leone and became Associate Professor. During his stay there (1959-1971), he conducted two major research projects, one on the diamond industry and another on Lebanese traders. Both projects resulted in the publication of books.

Preface

I had the good fortune to teach Economics at the University of Sierra Leone during the 1960s. All who lived and worked in West Africa at that time remember the political and economic optimism of those days. I shared and enjoyed this mood of buoyant expectation. I also benefited from its stimulating effect on academic research. It now seems so far away. When the wave of Afro-pessimism engulfed us in the 1980s, I had to ask myself whether the earlier optimism had been an illusion, if we had been collectively blind to the real facts. I am still struggling with the problem of doing justice to both the optimism of the 1960s and the recent pessimism. The best way to escape the mood of the moment seems to be the examination of a long period, as is done in this book.

A major objective of the new African universities of the 1960s was the Africanization of the curriculum. It was believed, on good grounds, that past teaching had been insufficiently adjusted to the situation in Africa and was therefore, to a large extent, irrelevant. I enjoyed the challenge of Africanizing my own teaching. This included, I felt, a willingness to think my lectures through from the supposition that Sierra Leone was the center of the world. These mental gymnastics turned out to be a first step in the writing of this book, which systematically adopts the African view—with African not in the social sense of the African peoples, but in the geographical sense of the African continent.

Africanizing the teaching program demanded an awareness of potential discrepancies between the theories developed in the West and the phenomena observed in Africa. In the field of economics, this was a special challenge because of a long-term tendency toward abstraction in this discipline. During the 1960s and 1970s, many Africanists felt that there was a poor fit between the economic theory they had learned and the facts they observed. But not everybody reacted to this discovery in the same way. Many scholars felt they had no choice but to reject neoclassical economics and to adopt an alternative theory (Dependency, Center-Periphery, Marxism, or neo-Marxism). I have not gone so far. My approach is that of modifying or extending classical economic theory. It must be added that the discrepancies I found were mainly spotted during interviews with businessmen in Sierra Leone, most of them in the commercial sector. I talked with Africans, Europeans, Lebanese, and Indians. In the interviews, I learned to value their insights and paid attention to their terminology, particularly if it deviated from what I had been used to.

I have many debts to acknowledge. First and foremost, to the people I met in Sierra Leone, Cameroon, Madagascar, Niger, and Kenya. Their interest and kindness have made me feel at home in Africa and have stimulated my research. I would like to specifically mention the businessmen. In the African setting, many interviews were more relaxed and personal than was usual in Europe.

Second, to my colleagues at Fourah Bay College in Freetown, in particular to Dr. N.A. Cox-George, my first head of department. My 12-year stay (1959 to 1971) at FBC, the oldest component part of the University of Sierra Leone, was a happy and inspiring one.

Third, to past and present colleagues at the African Studies Centre in Leiden. I want to mention five by name because they closely cooperated with me for shorter or longer periods: Cor Muntjewerff, Paul Hesp, Tjalling Dijkstra, Wim van Haaren, and John Houtkamp. Willem Veerman helped me with technical matters.

Fourth, to my colleagues at the Department of Marketing and Marketing Research at the Agricultural University in Wageningen. Among them, Aad van Tilburg deserves special mention.

The whole manuscript was read and commented on by Maja van der Laan-Bachofen, Cor Spruijtenburg, Aad van Tilburg, and Jennifer Wong. Particular chapters were read by Euan Fleming and Cor van der Plas. Of course, none of them is responsible for any errors of fact or interpretation in this book.

H. Laurens van der Laan
Leiden, Netherlands

Chapter 1

Introduction

This is by no means the first book on African export agriculture. The subject has given rise to a vast amount of literature including many thorough and competent studies. Then why would an economist like me want to add another one? More specifically, what deficiencies in current economic analysis could justify such an endeavor?

The main deficiency I see in the existing literature is an excessive reliance on market theory. Nearly all authors (not only the economists) take the world market as a starting point and suggest or postulate that the producers in Africa directly participate in it. But this does not fit the facts, for the great majority of these producers, both present and in the past, have been prevented from direct participation in the world market by its scale and complexity. In this book, I offer a new tool to study the actual and potential complications of indirect participation. This tool sheds new light on several other aspects of African export agriculture.

SEARCHING FOR A TOOL

My search for a suitable analytical tool involved three steps. The first one was the selection of Marketing Channel Analysis (MCA). MCA is discussed in detail in Chapter 2. Here, it suffices to say that the product is followed from origin to destination as it flows through the channel. Moreover, the actors, successively in possession of the product, are studied with regard to their objectives and operations. It is one of the special features of many African export crops that they undergo little or no transformation. It is therefore possible to identify the product all along the route, which encourages researchers to use MCA. Moreover, although an offshoot of economics, MCA has a different focus: it is not the market but the marketing channel that occupies the center of the stage. And whereas the market is assumed to be permanent, the marketing channel is seen as vulnerable. Its continuity is never taken for granted.

My second step was to identify a special class of marketing channels, the *international marketing channels* (IMC). A major advantage of the IMC is that it offers a new unit of analysis which for our purposes is superior to frequently used alternative units such as the country and the world economy. The former is unsatisfactory because it is too small and therefore incomplete. The latter is too large and unwieldy and permits only vague conclusions. In a vertical sense (i.e., in the direction of the product flow) an IMC is larger than the country because it is supranational. However, in the horizontal sense, it is narrower and thus smaller. I may add that, because of my economic training, I have found it difficult to break away from the country as the unit of analysis.

My third step was the demarcation and introduction of the *Trans-Oceanic Marketing Channel* (TOMC). The adjective trans-oceanic carries the notions of long distance and seaborne. (More will be said about this adjective in Chapter 2.) The TOMC is narrower in scope than the IMC in the same way that trans-oceanic trade is narrower than international trade: it excludes short distance trade and trade that depends on rail, road, or air transport.

The TOMC is a more effective tool than the IMC because of the following advantages. First, it is always possible to distinguish three sections in the channel: an *upper section* in the *country of shipment*, a *lower section* in the *country of destination*,[1] and a *middle or trans-oceanic section*. Second, the middle section is nearly always characterized by unbridled competition, partly because ships are not tied to transport lanes and partly because the oceans are extraterritorial and therefore largely free of state regulation. Third, the activities of trade and transport in the middle section are nearly always separated: exporters and shipowners are two distinct classes of actor. Fourth, the relationship between exporter and shipowner is more or less the same all over the world. This is the case since the second half of the nineteenth century when standard practices and documents were developed. Fifth, seaborne transport is relatively inexpensive and permits the export of bulky products, a category to which many primary commodities, including agricultural ones, belong. Sixth, since seaborne transport is inexpensive, it is hardly threatened by the competition of rail and road transport, at least when long distances are involved. Seventh, seaborne transport is slow. The *velocity of the product flow* in the middle section is low. This has significant consequences for the exporter in the field of financing and risk bearing.

It may seem that I took these three steps on the basis of deductive reasoning. In fact, it is the other way around. It was during my research in Africa that I discovered that it was helpful and often necessary to distinguish trans-oceanic from international trade. It was only later that I transferred this distinction to the field of marketing channels. The reader should be aware of this in order to understand the relation between analysis and empirical material in this book. If I had started with MCA, I would have collected evidence from all over the world. As it is, all the evidence I offer about how the actors in TOMCs operate comes from Tropical Africa, an area which I have been studying for many years.

In retrospect I do not think it strange that it is an Africanist who proposes the TOMC as a tool for analysis. The volume of trans-oceanic trade, taken as a proportion of international trade, is very high in Tropical Africa. Overland, airborne, and short-distance seaborne trade are correspondingly low. (They are proportionally lower than such trade in Western Europe, the Mediterranean Region, the Caribbean Basin, and Southeast Asia.) Tropical Africa is special in that nearly all its seaborne trade has been trans-oceanic trade. I estimate that, measured by volume, Tropical Africa's trans-oceanic trade was over 90 percent of its international trade around 1900; it still accounts for over two-thirds today. Although my knowledge of the foreign trade of Australia and Latin America is limited, I would think that the TOMC tool is also useful for these continents. As with Tropical Africa, intercontinental exports by rail and road are impossible for cargo. Before the 1960s, sea transport was the only realistic possibility for cargo. Since then, the exporter must choose between cheap sea and expensive air transport.

Applying the New Tool

In this book, the TOMC tool is applied to Tropical Africa for the reasons previously explained. The tool is further applied to agricultural products. Since a restriction was necessary for pragmatic reasons (i.e., to limit the amount of research), taking only agriculture seemed justified because agricultural products dominate exports in most African countries and have done so for many years. If the book is going to be relevant for African governments and instrumental for economic development, a focus on agriculture can be convincingly defended.

Figure 1.1 shows a diagram corresponding to the basic TOMC for Tropical Africa's export crops. The upper section is located in Africa, where the *growers* and the exporters operate and where the crops originate, and the arrow pointing to the right indicates the direction of the product flow in the channel. The lower section lies in the countries of destination outside Africa where the *importers* and consumers live. The ports are located at the borderlines between the sections. This diagram is the starting point for the argument developed in this book.

The diagram is drawn so as to suggest distance and spatial relationships, but we should recognize that it also represents a time interval, namely the interval between the moments of harvesting and consumption. The variable time also figures in the velocity of the product flow. (A major reason for excluding airborne exports is that the velocity of seaborne exports is considerably lower than that

FIGURE 1.1. The TOMC for African Export Crops

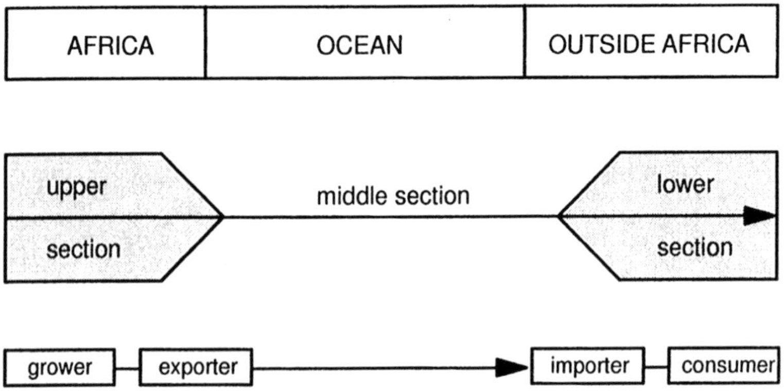

of airborne exports.) Seasonality, a natural feature of most of the crops under review, is another significant aspect of time.

The *exporter* in Africa is given prominence in this book. In what follows, exporters are consistently defined in terms of their location in the TOMC, on the borderline between the upper and middle sections. They are in possession of the trans-oceanic products in the port of shipment. This definition takes precedence over legal status (private or public enterprise) and nationality. (Many exporters in Africa have been foreigners.) Most exporters have to consider simultaneously *internal marketing* (in the upper section) and *external marketing*—two areas of analysis which have, unfortunately, drifted apart in the economic literature. Focusing on the exporter rather than the government introduces a strong bias toward microanalysis. This is indeed one of the significant points about this book. It adds microconcepts and data to the study of international trade and development economics—fields long dominated by macroanalysis.

ORGANIZATION OF THE BOOK

It is inherent to the TOMC approach that both a purely internal African and a purely external African perspective be ruled out. However, the way in which the two perspectives are combined varies. While in Parts I and III the external African perspective is strong, in Part II, covering two-thirds of the book, the internal African perspective predominates. This has a bearing on the combination of theory and empirical material. In Part II, the focus is on evidence and description. Elsewhere, theory plays a greater role: existing theory in Part I and new theory in Part III.

Part I (Chapters 2 to 5) begins with a chapter that elaborates the theoretical framework of Marketing Channel Analysis. Chapters 3 to 5 discuss how private enterprise created many new institutions for the middle section of the TOMCs, mainly in the second half of the nineteenth century. Four types of enterprise played a role: shipping, insurance, and telegraph companies, and international banks. Chapter 3 deals with shipping and Chapter 4 with the telegraph cables which were laid between Europe and Tropical Africa in the 1880s. The cables permitted the practice of forward selling which greatly altered the relationship between exporters and importers. Chapter 5 reviews how some trade associations helped to shape the world markets for trans-oceanic crops. At the end of Chapter 5, four types of crop with their corresponding world markets are identified: (1) *commodities*, traded on commodity exchanges; (2) *auction crops*, sold at periodic auctions; (3) *minor crops*, and (4) perishable crops. (For the last two only weak *network markets* exist.) This classification of crops and markets innovates and departs from more classic ones insofar as commercial and institutional aspects take precedence over production costs, prices, and demand characteristics. Finally, four types of TOMC are distinguished which determine the order of discussion in Chapters 7 to 13.

In Part II (Chapters 6 to 13), the TOMCs for Tropical Africa's export crops are described and analyzed. In Chapter 6, the three basic patterns found in the TOMC upper sections are examined. Chapters 7 to10 are devoted to the commodities and their TOMCs. Chapter 7 deals with external marketing. Chapters 8 and 9 describe how most exporters attempt to become channel leaders for the upper section. Chapter 10 investigates the actors' wish to increase the velocity of the product flow. Chapter 11 examines the minor crops. Chapter 12 deals with auction crops and recounts how auctions for coffee, tobacco, and cotton were introduced in East Africa in the 1930s—a fascinating commercial innovation. Chapter 13 deals with perishable crops and describes how the large-scale export of fresh bananas from five African countries started in the 1930s—another successful innovation.

In Part III (Chapters 14 and 15), the new findings of the TOMC approach are presented. In Chapter 14 I develop, for instance, the theory of exporter preference to explain the overwhelming role of the commodities in the African agricultural export baskets. This new theory has a bearing on the composition of the export basket in general, as well as on export diversification policies, not only in Africa, but also in other Third World countries. In the final chapter, the significance of my findings for economic theory and policy are examined.

The geographical scope of this book is wide. Tropical Africa roughly corresponds to sub-Saharan Africa, without the Republic of South Africa (see Maps 1 and 2). I include both anglophone and francophone countries, but the information on the countries where

Portuguese, Spanish, or Italian have been the official language is limited. The historical scope is a little over a century. (The choice of 1880 as the starting point will be explained in Chapter 4.) Periods of predominantly private enterprise are no less covered than those of the *export marketing boards* (EMBs). Of special interest are the periods of change: from private exporters to EMBs in the 1940s and 1950s, and in the opposite direction in the 1980s.

Assembling the Data

The data I have used consist of two broad categories: those on the TOMC sections and those on the actors. Data on the upper sections can be derived from the sectoral data in national statistics, such as those for the cocoa sector in Ghana. Sectoral data have been used by many authors and form the backbone of the descriptive literature. Unfortunately, sectoral data tend to stop short at the national boundary. Systematic information on the middle section is scarce because only a few organizations collect figures. Data on the lower sections vary from rich for some crops to poor for others. It is generally difficult to match upper- and lower-section data.

Most of the data on the actors are of a micro nature and were obtained from the actors themselves. If, among the actors in the upper sections, the exporters are the best source of information, this is because the average exporter has normally been larger and more sophisticated than, say, the farmers. I further gained much from studying proposals for sector reorganization, notably those in which the businessmen made a special effort to clarify their ideas.

The adoption of the TOMC as our unit of analysis has had serious disadvantages for data collection. Statistical documentation is poor, mainly because businessmen (our principal actors) collect fewer figures than governments. Moreover, the foreign trade statistics which are central when the country is used as the unit of analysis are of marginal use now: they rarely distinguish between transoceanic and other exports and, since they are collected for units of time determined by the calendar, they do not help to measure either the velocity of the product flow or specific time intervals. The evidence I can offer below is therefore patchy and more of a qualitative nature than is customary nowadays. Indeed, it would have been safer and easier for me to carry on from the existing theories, where the statistical abundance facilitates research! However, I contend that the TOMC approach provides gains in analytical precision that outweigh the loss in statistical documentation.

Assembling the evidence was an arduous task. As most of the questions I asked were new, the answers were either unknown or not easy to find.[2] As a result I had to consult and interpret a great deal of information; even so, the harvest of evidence was small. It would have been too small if I had not been able to cover a large area and a long historical period. As I charted my course in this extensive field of study (van der Laan, 1983a, 1986, 1987b, and 1993a), I found an unusual consistency in the historical and contemporary evidence that convinced me of the explanatory power of the TOMC approach.

GOVERNMENT'S ROLE

Clearly, the government is relegated to the second place in the analysis in this book. But since it remains important, the following points must be made.

First, I do not confine the term government to the independent African governments, as they have functioned since about 1960. I also apply the term to the colonial administrations of the past. True, the colonial officials were dependent on the metropolitan governments, but they had their own views and were not merely the mouthpiece of the minister for the colonies. In the same way, I use less developed country (LDC) both for the underdeveloped colonies of the past and the developing countries of the present.

Second, the role of governments tends to vary over time. Since the early 1980s, their role has been greatly reduced due to the adoption of structural adjustment programs, a decision in which donor countries and donor organizations have played their part. This is relevant for the policy recommendations in the final chapter.

Third, our focus is on the governments' relationship with the TOMCs for export crops. Rather than studying the extent of government participation in these TOMCs, I consider the underlying attitudes. We must remember that a positive attitude toward the TOMCs (as assumed in this book) is neither automatic nor constant. It is normally based on long-debated choices between (a) self-reliance and increasing exports, (b) inter-African trade and trans-oceanic trade (which are largely complementary), and (c) agriculture on the one hand and industry and mining on the other. In recent years the donors often stood on one side in these debates, while African organizations stood on the other. Thus, the Organization of African Unity with the *Lagos Plan of Action* (1980) and the United Nations Economic Commission for Africa with the *African Alternative Framework* (1989) endorsed self-reliance and inter-African trade.

Fourth, structural adjustment does not mean that governments blindly defer to businesspeople's decisions. By no means. Governments continue to give advice, to draft regulations, and to levy taxes— all of which affect businesspeople in the TOMCs. This book intends to help the policymakers to examine their attitudes. Are they consistent? Are the special characteristics of the TOMCs taken into account? Do the latter require modification of the simple models of (a) public versus private enterprise and (b) monopoly versus competition?

Fifth, the policymakers for whom this book is relevant form a large and varied group. Not only the officials in the ministry of agriculture are involved, but also those of several other ministries and the central bank, particularly when export agriculture earns large sums of foreign currency. The major donor organizations are also actively involved. In general, they have favored export-oriented development (World Bank, 1981). This book enables them to examine whether the policies they have recommended are propitious or need revision.

Glossary

auction crops: Trans-oceanic crops sold at periodic auctions. (For more detail see Chapter 5.)

commodities: Here defined as trans-oceanic agricultural products traded on commodity exchanges.

country of destination: The country to which the trans-oceanic product is shipped.

country of shipment: The country from which the trans-oceanic product is shipped.

export marketing board (EMB): A state trading enterprise for agricultural export products. An EMB is a public exporter. It usually possesses a buying and selling monopoly. (For more detail, see Chapter 7.)

exporter [in a TOMC]: The actor located on the borderline between the upper and the middle sections of a TOMC. The exporter is in possession of the trans-oceanic product in the port of shipment.

external marketing: The commercial activities related to the middle section and, in some cases, the lower section of a TOMC.

grower [in a TOMC]: The actor who grows or produces the product that flows through a particular trans-oceanic marketing channel. The term applies to farmers as well as plantation companies.

importer [in a TOMC]: The actor located on the borderline of the middle and lower sections of a TOMC. The importer is in possession of the trans-oceanic product in the port of destination.

internal marketing: All commercial activities related to the upper section of a TOMC.

international marketing channel [IMC]: A marketing channel located in more than one country. Since the product flow crosses an international border, it is possible to identify exports and imports.

lower section of a TOMC: The third, most downstream of the three sections of a TOMC. It is located in the country of destination.

middle or trans-oceanic section of a TOMC: The second of the three sections of a TOMC. Traders (i.e., exporters and importers) depend on ships and shipowners for the trans-oceanic transport of their products. Competition between shipowners tends to be strong.

minor crops: Trans-oceanic crops whose world market is only a weak network market. (For more detail, see Chapter 5.)

network market: A term coined to describe any market which lacks a public market place where buyers and sellers regularly meet face to face. It is the opposite of a physical market. (For more detail, see Chapter 5.)

Trans-Oceanic Marketing Channel: An international marketing channel whose distinguishing feature is that the product that flows through it is carried by ships across the ocean.

upper section of a TOMC: The first, most upstream of the three sections of a TOMC. It is located in the country of shipment.

velocity of the product flow: The rate at which the product flows through the entire channel and/or sections of it.

PART I:
THEORY AND GLOBAL ASPECTS

Chapter 2

Marketing Channel Analysis and the Basic TOMC

The study of marketing channels belongs to the field of economics, where it has carved out a special niche. Marketing channels may be defined as "sets of interdependent organizations involved in the process of making a product or service available for use or consumption" (Stern and El-Ansary, 1992, p. 1). The study of marketing channels has been greatly stimulated by the vigor of marketing management. Indeed, most channels do more than simply satisfy demand; they stimulate demand through promotional activities (ibid., p. 1). This explains why, until now, MCA has been applied mainly to the distribution of manufactured goods in Western economies.

Channel members are defined in terms of their function(s) in the marketing channel. The functional aspects in the definition take precedence over other elements such as the size and legal form of the organization. As a result, the terms channel member and actor—I prefer the latter—include both large and small enterprises. Moreover, the terms may refer either to an individual entepreneur or to a large incorporated enterprise or to a particular functional unit of such an enterprise. The reader should realize that the term actor, although it evokes the notion of an individual, normally refers to a collective and therefore impersonal entity. However, it is part of the functional approach that I attribute reasoning and decision making to this impersonal entity, as if it were a person.*

In the first part of this chapter, some MCA concepts are introduced; the second part elaborates the term trans-oceanic; and the

---

*For the sake of brevity, male pronouns are used in conjunction with these impersonal actors. This usage is not ideal because it fails to remind us of the role of women in many of these organizations.

implications for trans-oceanic marketing channels are explored in the final part.

MARKETING CHANNEL CONCEPTS

A basic concept is *channel sequence*. It refers to the order in which the economic actors figure in the channel—an order which is determined by the flow of the product. By following the product, an observer can identify the actors who are in possession of the product, all the way from producer to consumer. The observer is always able, for any pair of successive actors, to say that one is upstream and the other downstream. MCA tends to be most illuminating when the channel is long, i.e., when there are many successive actors.[1]

In many marketing channels, there are several successive markets. (Between every pair of successive actors there is a market.) For MCA scholars, the marketing channel as a whole is more important than those individual markets. Moreover, vertical relationships between actors (across the market)[2] are more important than horizontal ones (on the same side of the market). MCA sees markets as man-made—they may be created (by the insertion of an additional actor) and suppressed (by the elimination of another)—a view shared by the New Institutional Economics. Williamson (1975), for example, has shown that under certain circumstances *vertical integration* (i.e., the merger of two successive channel members, which, by definition, suppresses the intervening market), is a better solution than retaining the market—better not only for the enterprises involved, but also for the economy as a whole. Thus, the length of the channel, measured in terms of the number of actors and/or intervening markets, may vary over time.

The concept of sequence has also been used in other disciplines. First, geographers use the product flow concept. MCA may appear to them as just a fancy embellishment of product flow analysis—a field in which they have long done valuable work—including work on Africa. I hope to show that MCA offers more. Second, marketing chain scholars envisage a chain, along which the product flows from one stage or actor to the next. In fact, the words "chain" and "channel" are often used interchangeably. A recent development in this field focuses on global commodity chains (Gereffi and Korzeniewicz, 1994).

In every marketing channel, a number of commercial activities takes place. In Table 2.1, these commercial activities or functions are listed. (The list is longer than usual because in agricultural marketing and geography, it is common to include the physical functions of transportation and processing.) I have divided the functions into two groups. The functions of Group 1 are transaction or market related; those of Group 2 relate to the interval between transactions. While marketing costs arise in both groups, transaction costs, a significant concept of the New Institutional Economics, are a special feature of Group 1. Another reason to make this distinction is that MCA and market theory occasionally lead to different conclusions (Stern and El-Ansary, 1988, p. 260n). The distinction helps to trace the cause more easily.

In most marketing channels, an element of design can be found. One or more of the actors coordinate the channels in order to achieve a certain objective. When the terms *channel system* and *marketing system* are used below, this element of human design is implied. The objectives may vary. In channel systems that distribute an industrial product, we focus on the manufacturers' objectives. What are their goals? Do they want to raise their prices or their turnover? If only one actor is responsible for channel coordination, MCA speaks of a *channel leader*. The other actors in the system

TABLE 2.1. Commercial Functions in the Marketing Channel

Group 1 (related to a transaction)	Group 2 (related to an interval between two transactions)
1. buying	8. physical possession
2. selling	9. ownership
3. payment	10. storage
4. market intelligence	11. transportation
5. negotiation	12. standardization
6. promotion	13. processing
7. ordering	14. financing
	15. risk bearing

Adapted from Stern and El-Ansary (1992); Kotler (1994); and Kohls and Uhl (1990).

accept the channel leader's leadership, and relatively long-term mutual commitments between channel members consolidate the system. As most of them have a vested interest in the system's survival, they invest in its continuity.

In the first instance, a channel system must be studied in isolation. However, we must recognize that usually some form of competition exists between the channel system under study and parallel systems. Thus, an industrial channel leader may shape and use his channel system to get an edge over his rivals. As MCA has been mainly applied to the industrial sector, it has to be modified in this book to suit agricultural sectors where the channel includes an assembling or bulking stage.

MCA has further introduced the following contrasting concepts. A conventional marketing channel (CMC) is a "piecemeal coalition of independently owned and managed institutions, each of which is prompted by the profit motive but little concerned about what goes on before or after it in the distributive sequence" (ibid., p. 315). By contrast in a vertical marketing system (VMS), there is concern for the whole sequence. In a VMS, there is a "power locus . . . that provides for channel leadership, role specification, coordination, conflict management, and control" (ibid., p. 316). Coordination of a vertical system may vary from strong to weak. MCA scholars distinguish three basic types of VMS: corporate, contractual, and administered ones. In a corporate VMS all channel members are owned and operated by one organization; in a contractual VMS vertical coordination is accomplished through the use of contractual agreements; and in an administered VMS there is a kind of channel program that effects informal coordination (ibid., p. 348, p. 326, and p. 320).

THE ADJECTIVE "TRANS-OCEANIC"

In Chapter 1, several reasons for adopting the word "trans-oceanic" were given. If the adjective nevertheless seems strange and far-fetched, I would like to point out that it has the merit of linking the discussion of trade to the history of colonial expansion. Many historians acknowledge that the Imperialism of the later part of the nineteenth century was primarily an overseas (i.e., trans-oceanic) expansion of Europe.[3]

We must now consider the geographical implications of the adjective. In Map 2, Tropical Africa is divided into *Atlantic Ocean Africa* and *Indian Ocean Africa*, the former shipping across the Atlantic Ocean and the latter across the Indian Ocean. The dividing line—appropriate for most of the period covered in this book—is drawn along the political boundaries, with the exception of Zaïre. The eastern part of this country (shaded on the map) has experienced periods of eastward orientation toward the Indian Ocean and other periods of westward orientation. The same is true for Rwanda and Burundi.

The northern part of Atlantic Africa has always been called West Africa, with Nigeria (and sometimes Cameroon) taken as the most

Map 2. Tropical Africa's Maritime Orientation

 double orientation

MAP 1 Tropical Africa's maritime orientation

eastern territory. For the southern part of Atlantic Africa, there is, unfortunately, no standard name. Indian Ocean Africa includes East Africa, the Horn, the area formerly called Central Africa,[4] Madagascar, and some smaller islands.

In choosing between the terms "overseas" and "trans-oceanic," I considered the word overseas to be less precise for the inhabitants of the British Isles than *outre mer* is for the French and *Uebersee* for the Germans. The French and German terms suggest countries which are located not merely overseas, but in a different continent, which is an essential element in the analysis.

Generalizations about a large continent such as Africa are dangerous. The same holds true when using the word "trans-oceanic" to describe its trade. Exports to European destinations from northeastern Africa, i.e., from Djibouti, Port Sudan, and intermediate ports, are seaborne, but not trans-oceanic in the true sense. We ignore this nicety, however, and are partially justified in doing so since these ports had to ship across the Indian Ocean during periods when the Suez Canal was closed. Some inter-African trade is also seaborne; for instance, the shipping of kola nuts from Côte d'Ivoire to Senegal. The products in such trade are excluded from the analysis. Finally, the export statistics of the countries in southern Africa do not normally distinguish between export overland to South Africa and trans-oceanic export to Europe. This complicates the study of these exports.

Apart from trans-oceanic, which is a purely neutral descriptor, I use *ocean-straddling*, always with a positive connotation, to emphasize human endeavor; it is used to characterize organizations, contracts, markets, and relationships of mutual trust. Thus, in an ocean-straddling business organization, there are departments under unified management, but on either side of the ocean. In an ocean-straddling contract, the signatories are located on either side of the ocean, and so on.

THE TOMCs FOR EXPORT CROPS

We are now going to combine the trans-oceanic concept with the concepts of MCA. Before doing so, we must briefly note that MCA can be applied to imports as well. As with exports, a high proportion of Africa's imports moves in trans-oceanic channels. Indeed, many authors have drawn attention to the particular composition of Africa's foreign trade: imports consist largely of manufactured goods shipped from the developed countries (on the other side of the ocean), while exports predominantly consist of primary products.[5]

Typical of the TOMC for export crops (see Figure 1.1) are the two ports, one *port of shipment* and one *port of destination*. The ports divide the product flow into three sections. It is typical of the middle section that ownership and possession (see Table 2.1) have to be distinguished: the shipowner is in possession of the products, but does not own them.

Following the product en route is easiest if the product undergoes no processing at all, as is the case with bananas. This is also relatively easy with coffee, tea, and tobacco, but it is more difficult with cocoa, most oil seeds, cotton, and rubber, since manufacturing processes intervene in the lower section. Manufacturing in the upper section is usually restricted to elementary processing.

Place and Time Gaps

A marketing channel overcomes place and time gaps (Kotler, 1994, p. 527). With regard to the place gaps in a TOMC, we must think of the distance between the point of first purchase (or harvesting) and the port of shipment, the distance across the ocean, and the distance covered in the lower section. Here we can rely heavily on geographical studies, which also describe the means of transport and investigate transport costs. They point out, for instance, that since consignments carried by oceangoing ships are larger than those carried by other means of transport, there is a concentration of stocks in both the ports of shipment and the ports of destination.

There are several ways of looking at the time gaps in a TOMC. First, there is the entire interval corresponding to the channel as a whole. Second, there are three intervals associated with the sections of Figure 1.1. They may be called macroeconomic intervals. Third, there are the microeconomic intervals during which a particular actor is in possession of the product.

As the product moves from producer to consumer, or final user, it becomes more valuable because its utility of place and utility of time increase. The study of value addition may be approached in three ways. First, the value added in the channel as a whole is studied. Second, the value added in the three sections is studied separately. Because of the systematic recording of values in the ports, the value added during the ocean voyage is relatively easy to ascertain (see further Chapter 4). Third, the value added by each successive actor is studied. Here, precise figures are difficult to obtain. However, it helps to remember that the export value represents the value that is jointly earned by the actors in the upper section.

As a rule, the time gaps are not constant over the months because practically all agricultural production is seasonal. Seasonal peaks and troughs are normal in the TOMCs. In the ports, there are busy and slack months. And if elementary processing takes place in Africa, there are likewise busy and slack periods in the processing facilities. The seasonality of the product flow affects the TOMC in several ways. It bears on matters of infrastructure, capacity utilization, the calculation of costs per ton, and the allocation of the overhead costs of the idle months.

Specifying Individual TOMCs

The elementary physical unit in the middle section is a ship's consignment. This has repercussions in the upper section where the consignment is assembled, and in the lower section where it is broken up. In foreign trade statistics, the consignments are grouped together. Thus, TOMCs may be specified in terms of the crop and the country of origin; for example, the TOMC for cocoa from Ghana or the TOMC for sisal from Tanzania. In addition, it would be possible to specify the country of destination; for example, the TOMC for cocoa shipped from Ghana to Holland or the TOMC for Tanzanian sisal sold to Britain. Whether or not the destination should be specified depends on the extent of competition in the world market. If competition is strong, we may omit the country of destination because competition ensures that there is no great price difference between selling cocoa to a Dutch or a British importer. In the absence of competition, it is desirable to specify the country of destination (see Chapters 11 through 13).

In the TOMC approach, countries simply appear as countries of shipment or destination. Political scientists may feel that I pay too little attention to political aspects such as the colonial or independent status of a country. Macroeconomists may object that I relegate important economic policies, such as those concerning the liberalization of trade, to the background. Indeed, this is one of the consequences of the MCA approach, but it opens our eyes to topics that have been neglected by political scientists and economists such as the institutions of the middle section and the influence of exporters.

MCA is interested in channel continuity. In our examples (cocoa from Ghana and sisal from Tanzania), the continuity of the TOMCs is beyond question, but other situations are less clear-cut. When an exporter experiments with a new crop, there is no point in speaking of a channel until a fair number of consignments has been shipped. Firm expectations about continuing demand from one or more importers are necessary before a TOMC can be said to exist. Hence, intermittent and irregular exports constitute an indeterminate twilight zone not captured by MCA. On the other hand, the seasonality of exports and significant annual variation in volumes are not valid arguments against using the term TOMC.

The TOMCs of several African export crops have shown remarkable continuity over the last 100 years. However, these figures are only the starting point for MCA's real work. MCA examines the major actors and their functions with a view to discovering changes in these over time.

Collaboration Between Channel Members

Until now, four types of channel members have been identified: growers (producers), consumers, exporters, and importers. In many TOMCs, there are additional actors. There may be a retailer between the importer and the final consumer, and a *middleman* or *intermediary* between the producer and the exporter.[6] Sometimes there are also temporary actors. Clearly then, TOMCs will differ in the number of successive actors involved. This difference in *configuration of actors* precludes a rigorous term-by-term comparison.

While we will encounter conflicts between successive actors, collaboration is the norm. Since our terminology should be clear, I use *collaboration* for the joint efforts of successive actors, and *cooperation* for those of *parallel actors*. A telling metaphor to illustrate efficient and satisfactory collaboration is the relay run. Just as the athletes collaborate to carry the baton from start to finish, so do the successive actors collaborate to move the cocoa or cotton from producer to consumer.[7]

While the metaphor is helpful to study vertical relationships, it does not provide answers to the following questions. Does the number of actors increase or decrease over time? How much shifting and reshuffling of functions occur? How strongly are upstream and downstream markets geared to the international market? And so on. Here, it suffices to say that considerable changes in the configuration of the actors in the TOMC sometimes occur—changes which do not always show up in the foreign trade statistics.

While the TOMC approach probably appeals to the economist, the historian may object that it is not a neutral tool. A historian may rightly point out that nearly all TOMCs for African export crops were initiated from a downstream base and then gradually extended upstream. African farmers, for instance, could only participate on the terms set by the downstream actors. Moreover, for many decades, the exporters and some middlemen were non-Africans. Did not this cause a bias in the distribution of the total value added, and create a dependent relationship for the African upstream actors? These questions remind us of the issues of colonial exploitation, unequal exchange, and, more recently, of the objectives of the Fair Trade organizations. However, it is not the whole story.

I contend that such dependence has been counteracted by a process of *commercial emancipation* among the upstream actors. Their commercial knowledge and bargaining strength increased because they gradually came to understand the operations and problems of the next downstream actor. I focus on spontaneous emancipation. This differs from the popular view that the African actors cannot fend for themselves and have to be helped by the government. To document the process of emancipation, we must pay attention to the handicaps the upstream actors faced (language barriers, cultural gulfs, and limited access to commercial knowledge) and how and when these were overcome. It is also illuminating to look at the positions occupied by Africans, Europeans, and Asians (e.g., Lebanese and Indians) in each TOMC. The only shift of position that has been systematically studied is that of Africanization (i.e., of Africans replacing non-Africans), but other shifts also deserve to be studied. Fortunately for our purposes, earlier work frequently points to the race of growers, processors, transporters, middlemen, and exporters. Although such references to race are now considered inappropriate, some information from these older publications is quite useful.

Channel Coordination

Most of the TOMCs under review in this book are coordinated by markets rather than by channel leaders. Put differently, conventional marketing channels (CMC) are the rule. There are two reasons for this. First, since the TOMCs are long, the obstacles to unified control over the entire channel are very great. Second, shipping promotes international competition and thus strengthens the coordinating role of world markets. To suppress this market is virtually impossible for a channel leader, no matter how big he is.

Does this mean that Vertical Marketing Systems are absent? I coin a new term to make this question more precise. In a *Vertical Ocean-Straddling Marketing System* (VOSMS) a channel leader coordinates the activities of an exporter in Africa with those of one or more importers on the other side of the ocean. Such VOSMSs have not been entirely absent, but they have been rare, mainly because they require a special and expensive initiative.[8] We encounter a few examples in Chapters 11 to 13. In some of these, a Trans National Corporation (TNC) has played a significant role.

In the colonial period, VOSMSs had a better chance of success because the authorities promoted trade between the colonies and the metropolis. French colonial protection goes back to 1920 and British imperial preference was introduced in the early 1930s (Fieldhouse, 1971). Spain, Portugal, and Italy used similar measures.[9] After independence, the European Economic Community preserved some of these colonial elements by means of the conventions of Yaoundé and Lomé. In addition, many bilateral treaties between African and European countries were concluded concerning currencies, air and shipping lines, technical assistance, etc., but on the whole, the political basis for VOSMSs weakened after independence.

The statement that the overwhelming majority of the TOMCs are CMCs has to be modified. If the middle section resembles a CMC, it does not follow that the upper and lower sections form CMCs. In fact, there may be *partial vertical systems* in those sections. Thus, there may be *distributors* in the lower sections, enterprises with a strong brand name which enables them to control retail distribution. Such channel leaders are typical of the TOMCs for tobacco, tea, and to a lesser extent, coffee.

Partial vertical systems may also occur in the upper sections. In fact, they are common there if the trans-oceanic crop is grown by small farmers (see Chapters 8 and 9). In that case, the partial VMS is a collecting or assembling system, which resembles a river drainage system (Walker, 1959, p. 6) and is composed of all the channels between the farms in the hinterland and the port itself. It is in the ports of shipment—the mouth of the river in the metaphor—that the channel leaders emerge.[10] Since a collecting VMS is a convergent system, the exporter downstream is typically larger than the growers or intermediaries upstream. The number of parallel growers who sell to one exporter (or one intermediary) is a measure of the convergence of the channel system. A high *convergence factor*, which has been common in Tropical Africa, tends to (a) lead to severe competition among parallel actors and to (b) facilitate the emergence of a channel leader.[11]

Glossary

Atlantic Ocean Africa: That part of Tropical Africa whose trans-oceanic exports are shipped across the Atlantic Ocean.

channel leader: Actor who exercises control over other channel members and coordinates a marketing channel system. In this book, the term is also applied to an actor whose control is confined to one section of a TOMC.

channel marketing system: Here defined as a marketing system whose unity and coherence is at least partly the result of human design and effort.

channel sequence: The order in which the actors or channel members appear in the channel—an order based on the direction of the product flow.

collaboration: The joint efforts of successive channel members and the ways in which they work together.

commercial emancipation: A process by which some actors in the channel increase their countervailing power, notably their bargaining strength in transactions with other actors, either upstream or downstream.

configuration of actors [in a marketing channel]: An overall indication of the number, sequence, and relative size of the channel members.

convergence factor: A measure for the convergence of a marketing channel system. The number of growers selling to one intermediary indicates the degree of convergence at the intermediate level. The number of intermediaries selling to one exporter does the same at the export level.

cooperation: The joint efforts of parallel channel members and the ways in which they work together.

distributor: Here defined as an actor with a strong brand name who is able to have considerable control over retail distribution in the lower section of a TOMC. (For more detail, see Chapter 12.)

Indian Ocean Africa: That part of Tropical Africa whose trans-oceanic exports are shipped across the Indian Ocean.

middleman or intermediary: In this book, defined as the actor between the grower and the exporter.

ocean-straddling: Combining and coordinating activities on either side of an ocean.

parallel actors: Channel members who are located and operate at the same stage or level of the channel system. They do not normally trade with each other.

partial vertical system: Here defined as a vertical marketing system confined to either the upper or the lower section of a TOMC.

port of destination: The port in which the trans-oceanic journey of a product comes to an end.

port of shipment: The port in which the trans-oceanic journey of a product begins.

vertical integration: The merger of two or more successive channel members. Also, the situation which results from the merger.

Vertical Ocean-Straddling Marketing System (VOSMS): A vertical marketing system with a high degree of coordination across the ocean, that is, between one exporter and one or more importers.

Chapter 3

The Role of Shipping

This chapter deals with the oceangoing ships that have to be employed to carry the products that flow through the various TOMCs. In particular, the relationship between the *shipowner* and the traders is examined. Until well into the nineteenth century trans-oceanic trade and transport had been combined. It was only when transport was separated from trade that the exporters emerged as a special category of trader. At that time, most of the shipowners operated from Europe. In this chapter, Europe (with the importers) is contrasted with the other continents, where the exporters were domiciled. In the first or global part of this chapter, no distinction is made among these continents, but in the second part, the discussion is limited to Tropical Africa.

GLOBAL DEVELOPMENTS

A Historical Perspective

As trans-oceanic trade expanded in the nineteenth century, a process of specialization occurred, the essence of which was that the number of actors increased and that new forms of cooperation between them developed. In Figure 3.1, which illustrates this process, the physical product flow is kept the same in the three parts of the diagram; a certain quantity of a particular product is exported from a port of shipment to a port of destination in Europe.

In Phase 1, there was only one actor, the *trader-shipowner* in Europe. In Phase 2, there were three actors: one in the marketing channel (the marketing agent in Europe is dependent and therefore

FIGURE 3.1. Three Phases of Trans-Oceanic Trade

OUTSIDE EUROPE   OCEAN   EUROPE

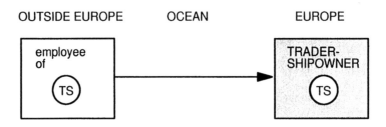

3.1a Phase 1. Trans-Oceanic Enterprise

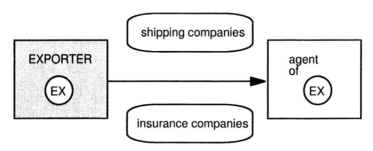

3.1b Phase 2. Trans-Oceanic Trust (consignment sales)

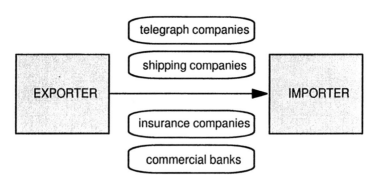

3.1c Phase 3. Trans-Oceanic Transactions (CIF trading, forward selling)

not counted and two *supporting enterprises* outside: the shipowner and the underwriter. It was a major step forward that in Phase 2 the activities of trade and transport were separated. There was also a geographical change. The principal actor was not domiciled in Europe, but in one of the other continents. In Phase 3, there were six actors: two within the channel and four supporting ones outside. Discussion of this phase must be postponed until the next chapter, but it is included in the diagram to illustrate the trend toward specialization.

Phase 2 followed Phase 1 in the middle of the nineteenth century in the wake of the new shipping lines. The establishment of the shipping lines was associated with technical changes: the steam engine replaced the sail (which permitted the introduction of dependable sailing schedules) and the steel hull replaced the wooden one (which permitted the construction of larger ships). Carrying costs and freight charges fell considerably—by an overall reduction of around two-thirds. The result was a rapid growth of seaborne trade, particularly that of long-distance trade. Trans-oceanic exports and imports soared. Phase 3 had to wait until around 1870 when the telegraph cables were installed. Neither 1850 nor 1870 can be taken as the end of an era because in both cases, the previous pattern continued to coexist with the new one for many years.

Phase 1

Phase 1 lasted for about three centuries, from the time of the Great Discoveries until the early nineteenth century. The trader-shipowner ran a *trans-oceanic enterprise*. He employed people, made all of the decisions, and bore all of the risks. In retrospect, we see how difficult his tasks of coordination and risk-bearing were. Apart from the great and well-known risk of loss at sea, there was the risk of a long voyage, caused by calms at sea or problems to fill the ship with a return cargo. There was the commercial risk that prices in Europe had fallen between the day the ship sailed and its return, and the lack of information due to the absence of contact between the captain and the trader-shipowner. In fact, the difference between a long voyage and a total loss could not be established unless and until the ship returned.

Phase 2

The relationships between the actors in Phase 2 were new in the sense that they had been absent in Phase 1. The exporter now entered into three contracts: a transport, an agency, and an insurance contract.

The transport contract referred to a *consignment*. (Normally there was one contract for each consignment.) The word "consignment" was appropriate because the exporters entrusted or consigned their products to the shipping companies for transport. In the literature on international trade, the commercial practice of Phase 2 is therefore usually referred to as that of *consignment sales*. Delegation and trust were indispensable elements. Since the transport contracts were increasingly standardized, the new market for ocean transportation was rather transparent. In fact, the shipping industry became a truly global industry, unified by pervasive competition.

It was important for an exporter to find a reliable marketing agent or broker in Europe. We should remember that in the nineteenth century, the agent in Europe and the principal overseas rarely met. This increased the significance of *trans-oceanic trust*. Sometimes, there was a formal partnership between principal and agent. Some partnerships were formed by family members or friends.

As the risk of loss at sea diminished, the insurance companies modernized their operations. Marine insurance, a new branch of the underwriting sector, offered separate policies to shipowners and exporters. The former were now covered against the risk of losing their ships, and the latter against that of losing their cargo. The remarkable point is that the demarcation of risks and responsibilities between shipowners, traders, and underwriters was standardized on a global scale within just a few decades. The standardization of practices and documents facilitated the collaboration between the various actors, even when they were domiciled in different continents.

Standardization also affected packaging. With general cargo ships transporting cargo of all types and sizes, stevedoring was a skilled and complicated job. Damage to goods was hard to prevent during loading, stowing, and unloading. It could be minimized by using special packaging, ideally made of a sturdy yet cheap material. The jute bag proved to be a satisfactory packing unit for agricultural products. It was introduced and universally adopted in the trans-oceanic transport of such products at a time when labor was cheap and mechanical handling virtually absent. The bag size was determined by human muscle power: a full bag could be lifted by two men and carried by one. Similarly, the weight units of cotton and tobacco bales, and the size of tea chests and casks for oil were determined by what dockers could handle. Almost automatically, these units were adopted in the upper sections of the TOMCs.

SHIPPING LINES

The new shipowners could choose between three alternative strategies for their vessels: they could operate them as tramp ships, they could offer them for charter, or they could use them to set up a shipping line which operated as a public or common carrier. The last option was the most ambitious, but also seemed most advantageous. The shipping lines[1] had to set up a shore organization with agents in all ports of call, at home and overseas. These agents provided information on sailing schedules and freight rates. They also acted on behalf of the shipowner in accepting cargo.

The shipping lines signified the modernization of the shipping industry.[2] The shipowners that operated a line strove to establish a reputation for regular, dependable sailing schedules, short intervals between successive sailings, and honest and competent personnel. This permitted, for instance, the upgrading of the transport contract. At first it had concerned only the two contracting parties. In due time, other enterprises, notably commercial banks and insurance companies, began to attach value to the bill of lading, as the contract document came to be called.

It was only one step further to view the bill of lading as a document giving the holder legal title to the consignment. For a bill of lading to become negotiable, it had to contain a satisfactory description of the consignment—satisfactory in the sense that embezzlement was impossible and that in case of loss at sea, the financial loss could be accurately established in order to claim insurance. Since the negotiability of the bill of lading depended on the shipowner's reputation, exporters preferred to deal with reputable shipowners.

The separation of legal title of ownership from possession, accomplished by the bill of lading, was later duplicated on shore. Some businesspeople in the ports of destination set up specialized warehouses where trans-oceanic crops could be stored under technically ideal conditions. They issued warehouse receipts to traders who deposited a consignment of their goods with them. These documents, once standardized, greatly facilitated trade because legal title could be transferred, while the corresponding physical stocks remained in the warehouse. The receipt thus became negotiable and, consequently, an instrument for obtaining bank credit. Both the shipowners and the warehouse owners had to make clear that they were not interested in buying the products. In this way they preserved a reputation for being reliable custodians.[3]

Most shipping lines operated a cargo reservation system. An exporter could contact the line's agent and reserve space for his cargo. Both the ship and the day of shipment were specified. Although there was no charge for these reservations, they nevertheless had the nature of contracts, distinct from the transport contracts. They were particularly valuable for the exporters of agricultural products because the transport requirements of those products were seasonal. To sum up, a shipping line rendered three services: it carried goods, it issued negotiable documents, and it operated a cargo reservation system. While the first service was by far the most important, the other two should not be overlooked.

To say that the reservations were free of charge, is in a way misleading. In return for the continuity they provided (on which the reservation system rested), the lines expected loyalty from the exporters. One instrument to ensure loyalty was the rebate, often a *deferred rebate*, which could amount to 10 percent.[4] Exporters forfeited their rebate if they shipped some of their consignments with an outside ship. Outside shipowners could quote a much lower freight rate, but an established exporter would normally refuse them; because even substantial savings on several consignments would not make up for the loss of the rebate. The rebate system was effective in tying clients to a shipping line and in restricting competition on the line's route.

Conflicts Between Exporters and Shipowners

If conflicts arise between exporters and shipping lines, they normally concern the level of the freight rates. The exporters may complain that the rates set by the line are too high. Since information about rates in the world charter market can easily be obtained, most exporters are able to compare them with what they themselves have to pay. Charter rates exhibit large fluctuations, and exporters tend to become restive when rates are low. Usually, exporters do not go farther than complaining. It is only in exceptional cases that they decide to charter their own ships. The opportunities to do so are most propitious when the exporter needs transport for large volumes and is also involved in the import business. We will come back to this point in the section on Africa.

Another area of conflict has been the rebate system. Exporters want a rate system without rebates in order to be free to switch from one shipowner to another or to combine the services of several shipowners. Exporters also complain when the deferred rebate is paid late. (In the past, some lines paid after a delay of up to one year—hence the term deferred—which meant that the shipper provided some short-term financing to the shipowner.) A further ground for complaint is the fact that the line may not grant a rebate for all products.

It would be wrong to think that freight rates and rebates are the only bones of contention between exporters and shipowners. Other causes of conflict come to light when we briefly review various wishes of exporters and shipowners. We will first consider those of the shipowners. Shipowners want a high load factor, that is, their ships should be as full as possible because this is a major element in profitability.[5] If possible, they want a high load factor in either direction. (This objective is relevant for general cargo ships only; for special cargo ships, see below.) Sometimes the line offers concessional rates for certain export products to increase the load factor. This may lead to complaints from exporters who pay normal rates for comparable products.

We now turn to the exporters' wishes. As exporters are time conscious (as explained in Chapter 10), they want to reduce two intervals: the interval during which their products await shipment and the interval that they are on board. The first interval may be caused by temporary discrepancies between demand for, and supply of shipping space. As we saw, exporters of agricultural products demand shipping space in a relatively short season. Is the shipping company prepared to adjust its shipping space to agricultural exporters' demands? Here we see the significance of (a) the reservation system in helping the shipping line to know future demand and (b) chartering as a means of adjusting capacity. Is the line prepared to charter ships to cope with the seasonal bulge in cargo? Is it prepared to do so even when charter rates are high so that the voyage under charter leads to a loss?[6]

The second interval concerns stocks on board. It is illuminating to say that consignments are stored in a ship and thus represent stocks for which the exporter has to provide finance in the same way as for the stocks on shore. Clearly, the need to borrow money is less when the period on board is shortened. The shipping line is in a position to shorten this period in two ways: by operating ships with a higher service speed and by reducing the number of ports of call on a route. The first point is obvious, but the second needs further explanation. In each port, some time is needed for berthing and (un)loading. For the cargo on board, this is unproductive time. The owner's costs continue, but the cargo does not get any nearer to its destination. Exporters at the end of the line are particularly inclined to complain about the number of intermediate ports. They prefer direct sailings to the port of destination. The shipping line may oppose this solution if it lowers the load factor.

Some exporters ask for special ships, that is, ships constructed to carry special cargo in bulk. Thus, liquid cargo is best transported in tankers and ore in bulk carriers.[7] The technical advantage of such special ships over general cargo ships is that loading and unloading take less time because the cargo is homogeneous. The disadvantage is that no cargo is carried on the return voyage so the ship sails under ballast half the time.

On shore, special facilities are usually necessary for handling the special cargoes. Thus, the loading and unloading of ore requires special installations. Similarly, storage tanks and pumps on shore are needed to complement tankers. In general, such installations were first established in the ports of destination.

In agriculture, special cargoes are rare. The main categories are vegetable oils and perishable crops. In the first category tankers reduce carrying costs considerably. However, the shipping company is unwilling to employ special ships when it anticipates a low load factor. This leads to disagreement with the exporters who say that special ships are absolutely necessary to keep their product competitive. As a compromise, the shipping company may modify some of its general cargo ships, equipping one hold for vegetable oil.

Refrigerated ships are required to carry perishable crops. In this trade, exporters and shipowners have always had to collaborate. A kind of prior agreement between them was often necessary before loading facilities were built in a port of shipment. Such agreements reduced competition between shipowners and exporters. Indeed, shipowners, although not members of the marketing channel, affect the degree of competition within the world market (see Chapter 13).

Finally, we should note that the wishes of exporters and shipowners fully agree on one point: they both want to prevent port congestion. They jointly ask governments to provide large, well-equipped ports.[8] Exporters and shipowners are also in favor of ample warehouse space because it facilitates the preparation of individual consignments before the ship arrives and so reduces the time the ship spends in port.

DEVELOPMENTS RELEVANT
TO TROPICAL AFRICA

The colonial expansion of the 1880s encouraged European shipowners to set up or improve shipping lines between the metropolitan country and the new colonies in Africa. In organizing these lines, European shipowners were able to use the practices and documents developed elsewhere. The shape of the African continent suggested two major routes to these shipowners: one serving Atlantic Africa, and the other serving Indian Ocean Africa (see Map 2, p. 19). As we saw in Chapter 2, Atlantic Africa consisted of two areas: West Africa and the southern part of Atlantic Africa. Shipping to and from West Africa will be discussed in some detail below, but first we will make some brief remarks about the other areas.

The situation in Indian Ocean Africa was more complicated than in Atlantic Africa. First of all, the longer voyage had serious consequences for costs. Second, there were fewer countries and fewer ports, which made the planning of shipping lines easier for the shipowners, but also made it more difficult for them to earn enough profit. Third, Mozambique alone could not easily justify a separate Portuguese line, nor Madagascar a French line, nor pre-1914 Tanganyika a German line. Fourth, the Suez Canal, opened in 1869, had improved the economic position of Indian Ocean Africa considerably, but the fees charged for passage through it raised transport costs. Moreover, in times of war, the Canal could be closed, forcing ships to sail round the Cape.[9] Fifth, destinations outside Europe, notably India, and later Japan, were attractive.

The southern part of Atlantic Africa presented a dilemma to shipowners. One possibility was that they saw the area as an appendix of West Africa and set up a line for both areas together. The disadvantage of this solution was that there were too many ports. Unless ships bypassed some ports, the round voyage would be too long. The alternative was that of direct sailings. This policy was, for instance, chosen by the Belgian line.[10]

West Africa enjoyed a privileged location because of its relatively short distance to Europe. Moreover, the area was productive and accounted for a large part of Africa's exports. It was further important that the volumes of northbound and southbound cargo did not greatly differ, which was attractive for the shipowners. The fact that West Africa was a promising area for the shipowners may help to explain why they restricted competition among themselves by means of a so-called *shipping conference*. The West African Conference was established in the 1890s when Woermann (a German line) joined Elder Dempster from Liverpool.[11] For Elder Dempster, Nigeria was the terminus[12] and for Woermann, it was Cameroon.

Conflicts and Disagreements in West Africa

Much has been written about the West African lines (Leubuscher, 1963; Davies, 1973; and Fieldhouse, 1994). Here I select a few points that shed light on the relationship between the exporters and the lines. There were always latent disagreements about freight rates. Disagreements tended to come into the open when an economic crisis caused agricultural prices to fall. A dramatic conflict occurred in 1929. On one side stood the United Africa Company (UAC), the result of a merger of several trading companies earlier that year. UAC was a giant trading company in the West African context. On the other side stood the West African Lines Conference, with Elder Dempster as its leader. In late 1929, the negotiations between UAC and the West African Lines Conference broke down. UAC had demanded better terms, notably lower freight rates.[13] When the conference proved unbending, UAC announced that henceforth it would carry its own cargo in charter ships and no longer use conference vessels.[14] UAC followed this policy for the next 20 years.[15]

Several factors played a role in the conflict in 1929. First, only a very large trading company had a chance of success when it came in conflict with the West African Lines Conference. In this connection, the effect of the shipping lines on the scale of the export function should be mentioned. In general, the lines offered good opportunities for small-scale exporters.[16] Large exporters were less happy with the lines, especially if they were unable to obtain special rates. Second, UAC, like most exporters in West Africa, was an export-import company. It had both north- and southbound cargoes. Apparently, the imbalance between the two was not too great; with proper management, UAC could secure a high load factor in both directions.[17] Third, the possibility of direct sailings from West Africa to Europe was now open to UAC. For cargoes to and from Nigeria, the principal area of UAC's operations, the interval on board could thus be reduced.[18]

Another disagreement between exporters and shipowners concerned special cargo. Early in the 1920s, it was clear that the carriage of palm oil had to be modernized. Transporting it in casks or drums—the system then in use—was expensive, lowering prices paid to producers and discouraging production. Modernization of port facilities began in the United States, which bought a large part of West Africa's palm oil, and then spread to Nigeria, where one exporter, the Niger Company, built bulk installations in three ports (Burutu, Apapa, and Port Harcourt). It seems that Elder Dempster was slow in equipping its general cargo ships with tank holds for carrying palm oil in bulk. By the late 1920s, the capacity of the shore installations exceeded that of the ships, thus preventing the Niger Company from reaping the economies of scale of bulk handling.[19]

It seems that the general cargo lines were also reluctant to introduce refrigerated ships when, around 1930, the export of bananas from Tropical Africa began. At first, they equipped some of their general cargo ships with one refrigerated hold. It was only when this solution proved inefficient that they bought or ordered special banana carriers for the African routes (see Chapter 13).[20]

Glossary

consignment [in the shipping trade]: A quantity of a ship's cargo covered by one shipping document.

consignment sales: Here defined as the exporter's practice of (a) shipping a quantity of his goods to an agent on the other side of the ocean and (b) instructing him to sell the goods on his behalf.

deferred rebate: A rebate retrospectively paid by a shipping line to exporters that have shipped their cargo with this line during a particular period. Since the rebate is conditional on an exclusive relationship with this line (and possibly associated lines), it encourages exporters to remain with this line.

shipowner [in a TOMC]: The actor who owns the oceangoing ships that carry the products that flow through the TOMC. Normally, a shipowner is a company that owns several vessels.

shipping conference: An association of shipping lines serving the same region. The association is based on an agreement with regard to sailing schedules and freight rates, which removes or reduces competition among the lines.

supporting enterprises: Here defined as enterprises outside the TOMC which provide services and institutional safeguards to exporters and importers so as to reduce the risks and problems associated with their trans-oceanic transactions.

trader-shipowner: An entrepreneur who is both a trader and a ship-owner.

trans-oceanic enterprise: An enterprise that combines operations at home and on the other side of an ocean.

trans-oceanic trust: A relationship of trust between two channel members who reside on different sides of an ocean.

Chapter 4

The Role of the Telegraph Cables

Information travels faster than goods. This observation is central to this chapter. In our era of satellite telephones, it takes a strong mental effort to picture the time when information in trans-oceanic trade traveled as slowly as ships. Yet it was only 150 years ago that information, goods, and passengers all traveled at the same speed— the speed of the vessels that carried them. The change in information speed came when submarine telegraph cables were installed.

GLOBAL DEVELOPMENTS

Submarine Telegraph Cables

The technology of overland telegraphy was mastered in the first half of the nineteenth century, but manufacture of a submarine cable suitable for laying on the sea bed proved to be difficult. Success was first achieved in 1851 when a cable from Dover to Calais was installed. As the technical problems increased more than proportionally in relation to the distance to be covered, the laying of a trans-oceanic cable was a major challenge. The first cable across the Atlantic (after an unsuccessful attempt in 1858) was laid in 1866. This was followed by a cable from Europe to Bombay and the Far East in 1870 and a branch cable to Australia in 1871. A line from Europe to South America (crossing the Southern Atlantic and surfacing at Pernambuco in Brazil) was laid in 1874. Africa was the last continent to be linked to Europe. This happened in the 1880s (for details, see below).

The trans-oceanic cables were owned and exploited by private enterprise, but governments were involved in several ways. First, most governments were significant customers. Second, some colonial powers granted subsidies for those cable sections that were uneconomic, but nevertheless valuable for political reasons. Third, governments could grant (or refuse) the right to build a shore station where the cable surfaced. Because of these factors, it is useful to attribute a nationality to a particular cable. As a rule, the first trans-oceanic cables were British. Later, there were also French and German cables.

Cable-Based Ocean-Straddling Organizations

The new cables changed the character of the colonial empires. First, up-to-date information from the other side of the ocean strengthened the colonial administration. Supplies from Europe could be requisitioned more quickly and their deployment could be planned before they arrived. Military planning was strengthened. Second, the cables altered the relationship between the governor in the colony and the minister in the metropolitan capital. Formerly, the governor of a colony was a man with a large degree of discretion. (Trustworthiness was a major criterion in his selection.) After the cables were installed, the autonomy of the governor was reduced because he was expected to refer to his minister far more often. The old excuse for acting independently—that the situation would have gone out of hand if he had waited for instructions—was now rarely accepted.

The points just made were not only relevant for the colonial administration, but applied to all ocean-straddling organizations. Their effectiveness greatly improved, but there was a shift in power from the man abroad to his boss in Europe. Moreover, the relationship of trust between the two was replaced by one of command.

The effects of the cables on the shipping lines need to be specially mentioned. The formerly independent agents in the various ports were now forged into a well-coordinated enterprise, in which information flowed freely and quickly. Agents now knew how much room was still open in the holds of the ship before it arrived in port. The directors also gave agents limits for accepting reservations from exporters. In this way, the directors could achieve a high load factor for the ships. In fact, the modern shipping lines (for cargo and passengers) were inconceivable without the cables.

A number of European banks specializing in international trade responded to the new cables by opening branches abroad. The main services these banks provided for exporters were finance and guarantees for full and prompt payment by importers. To do this effectively, the banks had to rely heavily on the telegraph cables.

TRANS-OCEANIC TRANSACTIONS

Before the cables were installed, exporters depended on the practice of consignment sales described in the previous chapter. The new cables ushered in the practice of *trans-oceanic transactions*. The essential elements of this practice are shown in Figure 3.1c (see Chapter 3). The importer is now an independent actor who buys directly from the exporter. The fact that these two actors are several thousands of miles apart demanded some adjustments. Thus, there was no longer a sales contract signed by the two parties. Instead, there were two cablegrams, one from either party, which together formed the legal basis of the transaction.

Figure 3.1c also shows the four supporting actors. The reliability of all of them—long taken for granted since then—was a precondition for the success of the new practice. First, the shipping companies had to be reliable in the sense that they adhered to their sailing schedules so that the buyer-importers could estimate with reasonable accuracy the date of arrival of the consignment they had bought. Second, the telegraph companies had to provide reliable, continuous service to exporters and importers. Moreover, the cablegrams had to be acceptable as legal documents in case of problems of contract enforcement. Third, reliable banks guaranteed payment by the importer. Fourth, reliable marine insurance companies provided coverage against the risks associated with the transportation of the contracted consignment. In addition, the description of the quality had to be brief (to prevent an expensive cablegram) but clear, that is, unambiguous in case of a later conflict. (See Chapter 5 on grades.)

The reader will look in vain for the term trans-oceanic transactions in the literature. Instead, he/she will find *CIF trading*. Under a CIF export contract, which is a sales and shipment contract, the exporter pays the CIF (cost, insurance, freight) expenses and correspondingly increases the invoice. The importer does not have to bother about these expenses, when the consignment arrives at the CIF point in the port of destination. The FOB (free-on-board) contracts are a (less common) variant of the CIF contracts. Under FOB terms it is the importer who pays for the cost, insurance, and freight of the journey. The abstract terms CIF point and *FOB point* may be used as equivalents of port of destination and port of shipment respectively.

Customs officials in the countries of destination work with the CIF prices, that is, the prices agreed upon in CIF contracts, because those prices reflect the value of the consignment when it arrives in their country. Custom officials in the countries of shipment prefer FOB prices because they correspond to the value of the consignment when it leaves the country.[1] The difference between the FOB and the CIF value cannot be ignored. For many consignments, the CIF value is 5 to 10 percent higher.

A CIF contract is automatically a forward contract because the selling of the goods takes place before their delivery. Compared with consignment sales, the moment of sale is brought forward. While previously the sale of a particular consignment had to wait until it had arrived in the port of destination, now it took place before shipment. The gain in time was at least as long as the ocean voyage, perhaps, three weeks. In fact, the commercial interval could be much longer than the transport interval, if a sale several weeks before shipment suited both exporter and importer. This possibility, discussed in Chapter 7, is a reason for me to use the term *forward selling* rather than CIF trading.

As a rule, forward selling has no effect on the moment of payment. Payment by the importer is not due until the consignment has been delivered. Thus, although the cables enabled the exporter to bring forward the moment of sale, the moment of payment remained more or less the same as that under consignment sales. Exporters therefore badly needed credit to tide themselves over during the interval between the date of shipment and that of payment by the importer. The banks stepped in to solve this problem. Banks were willing to provide short-term credit for this interval. The basic document they required was the invoice which the exporter made out for the importer on the basis of the forward contract. The invoice specified the product, the product's price, the volume of the consignment, and the ship that would carry the product. This invoice, in conjunction with the bill of lading, was presented by the exporter to a bank in the port of shipment to support an application for credit. Thus, the burden of finance could be partially shifted from the exporters to the banks. The details of such financing are explained in many books and need not concern us here.

The Adoption of Forward Selling

The main point of this chapter is that the submarine telegraph cables created the opportunity for exporters to replace consignment sales by forward sales. The new practice was adopted on a global scale, but many questions remain as yet unanswered. Was the new practice adopted at once or only after the banks had established themselves abroad and had begun to extend credit to exporters? Was the switchover general, or did some exporters continue with consignment sales? And were some products more suitable for forward sales and others for consignment sales? Did periods of impending scarcity and rising prices favor the adoption? Perhaps the historical documents will yield more information when researchers become aware of these questions.

In any case, the telegraph cables were a major factor in incorporating the other continents into the world economy. For statistical analysis, it is further important to note that the quality of the export statistics improved after the changeover to forward selling because the figures on the customs declarations were now more firmly based. Under consignment sales, the export prices declared by the exporter were merely estimates of the final sale value; they reflected the exporter's optimism or pessimism rather than the final value to be determined several weeks later. Under forward selling, the figure on the customs form showed the FOB price derived from a definite CIF contract.

We should be clear about another element of the changeover to forward sales. Just as consignment sales were based on simple, personal trust, forward selling depended on institutional safeguards provided by the supporting enterprises shown in Figure 3.1c. There is general consensus about three component parts of this supporting complex: the banks and the shipping and insurance companies are always mentioned. However, the telegraph companies are generally omitted. It is unfortunate when their significance is overlooked.[2]

At first, the companies of the supporting complex were European (British, French, Belgian, etc.) because their headquarters were in Europe and the initiatives had come from there. In the other continents where only the branches operated, the complex was seen as foreign. This prompted a great deal of concern and envy. As a rule, commissions and fees made up most of the earnings of the companies in the complex. Since the services were essential, the companies could easily counter any opposition to new, higher fees by arguing that the continuity of their services was otherwise in jeopardy.

A Reallocation of Risks and the Role of Reputation

The changeover to forward sales brought a substantial reallocation of risks in trans-oceanic trade. Formerly, under consignment sales, many risks were borne by the exporter in Africa. Exporters consigned their products to the shipowners and to their marketing agents in Europe in the hope that they were competent, honest, and would look after their interests. However, there was the risk that their trust was misplaced—a risk that the exporter could not cover. Conversely, the buyers in Europe ran few risks. They could delay their buying decisions until they had seen the products.

The practice of forward selling increased the risks of the importer-buyers and reduced those of the exporters correspondingly. As a rule, the importers faced three risks of default. First, the exporters might be dishonest and not ship the consignment they had sold. Second, they might not be punctual and ship their products later than agreed upon. Third, they might be incompetent and ship goods of inferior quality, i.e., lower quality than specified in the contract. The first two risks came to an end when an importer heard from the shipowner that the consignment had been loaded. The third risk would continue to worry the importer until the consignment was unloaded and inspected.

No doubt, this reallocation of risks had an effect on prices, that is, the forward-selling exporters received a lower price than their conservative colleagues who continued to make consignment sales. This discount was largely compensated for by the advantage of an early sale, which brought earlier certainty for the exporters, not only piecemeal, i.e., for each consignment, but also for the whole season. Early certainty about sales simplified management and increased their credit standing with the banks. On the other side of the ocean, many importers were prepared to assume a larger part of the risks provided that this lowered the price they had to pay.

Under the new conditions, it was important for forward-selling exporters to build up a reputation for honesty, punctuality, logistic control, and consistent grading of their products. Since such a reputation could not be built up overnight, the exporters valued the epithet "long-established." A good reputation gave an exporter more leverage in negotiations and tended to raise the contract price.[3] Exporters were aware of this and made great efforts to build up and maintain a good reputation. These efforts fitted well with the trend toward larger commercial units. In considering the issue of *exporter's reputation,* we should realize that the average importer in Europe could not easily verify the standing of an exporter in one of the other continents. This problem was less serious when the exporter was linked to a reputable company in the importer's country—a factor which favored European exporters. Commercial banks were also a help because they used objective standards to assess the exporter's financial standing, which could serve as a proxy for reputation.

The practice of forward selling diminished the risks of the exporters, but did not fully eliminate them. Standard procedures were developed to cover these remaining risks. First, the risk of default by an importer (not paying after having taken delivery of the consignment) was taken care of by the banks. Second and less conspicuous, if prices had drastically fallen after the contract had been entered into, an importer might refuse the consignment by claiming (correctly or not) that the quality of the shipment was inferior to that specified in the contract. Objective standards of quality and arbitration procedures (see Chapter 5) took care of this risk. Third, an importer could also claim underweight or damage. Independent verification companies employing sworn inspectors in the ports of destination could be contracted by an exporter to counter this risk.

The relationship between exporters and importers, as shaped by forward selling, has remained the same in spite of later changes in telecommunication technology. The telephone facilitated negotiation and the telex offered direct communication from office to office, but the basic elements of the trans-oceanic transactions have not changed.[4]

DEVELOPMENTS RELEVANT
TO TROPICAL AFRICA

As previously mentioned, Africa was the last continent to be connected to Europe by cable—an indication of "the relative unimportance of that continent in terms of international economic life" (Latham, 1978, p. 35). The costs of the cables serving Africa were relatively low because they were constructed as branches of the intercontinental cables to Asia and South America. Thus, the cable in East Africa started at Aden (on the cable to Asia) and the main cable in West Africa started on São Vicente in the Cape Verde Islands (on the cable to Brazil and the remainder of Latin America).

The details are listed in Table 4.1 and shown on Map 3. It is interesting that, considering British-initiated cables only, East Africa was six years ahead of West Africa, largely because it provided a shorter connection to South Africa. In 1889, when the gap between Luanda and Cape Town was closed, the African continent

TABLE 4.1. Early Submarine Telegraph Cables Serving Sub-Saharan Africa

Intercontinental Cables	
1870	Suez - Aden - Bombay
1874	Lisbon - Madeira - São Vicente - Pernambuco
1884	Cadiz - Canary Islands - St. Louis - Dakar
African Cables	
1880	Aden - Zanzibar - Mozambique - Delagoa Bay - Durban
1885	Dakar - Conakry
1886	São Vicente - Bathurst - Freetown - Accra - Lagos - Brass - Bonny
1886	Grand Bassam - Cotonou - Libreville
1886	Luanda - São Tomé - Principe - Bonny
1889	Luanda - Benguela - Mossamedes - Cape Town

Sources: Bright, (1898); Lesage, (1915); Latham (1978).

MAP 3. Early Submarine Telegraph Cables Serving Africa

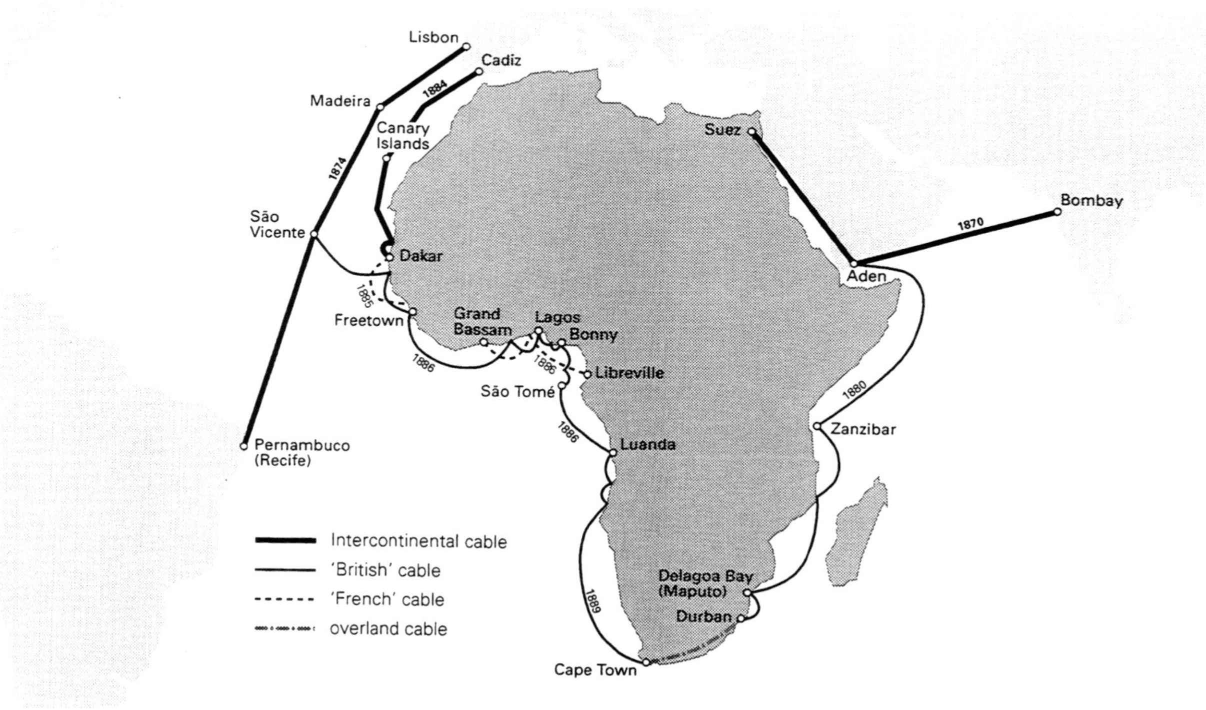

Bombay

1870

Aden

Suez

Zanzibar

1880

Delagoa Bay
(Maputo)

Durban

Libreville

Luanda

Lagos
Bonny

1886

1886

Cape Town

1889

Grand
Bassam

São Tomé

1886

Lisbon

Cadiz

1884

Madeira

Canary
Islands

1874

Dakar

1885

Freetown

1886

São
Vicente

Pernambuco
(Recife)

Intercontinental cable

'British' cable

'French' cable

overland cable

51

was encircled. It was in that year that the three telegraph companies operating in sub-Saharan Africa (the Eastern and Southern, the West African, and the African Direct) decided to operate in future as "one working system" (Bright, 1898, pp. 134–135).

None of the authors whose works I consulted drew attention to the fact that the cables were laid in the decade of the scramble for Africa. (The cable to Dakar was probably being laid while the Berlin conference was in session.) Obviously, the cables were important for colonial expansion. One aspiring colonial power, Belgium, was in a highly dependent position,[5] while another small country, Portugal, was relatively strong because of the shore stations on its territory. France was independent in West Africa for only a few years. Germany, dependent on British cables both in East and West Africa, finally decided to invest in a cable of its own, which was to serve both South America and West Africa. Monrovia was chosen as the African shore station, as the distance from there to Brazil was shortest. The German cable reached Monrovia in 1910 and Lomé in German Togo and Duala in German Kamerun in 1913.[6]

There is only indirect evidence of the effect of the cables. The international banks began to open branches in Tropical Africa in the 1890s.[7] There was further a trend toward larger trading companies in West Africa (Hopkins, 1973, p. 202) which may have been partly inspired by the companies' wish to engage in forward selling.

In the general part of this chapter, I drew attention to the efforts of exporters to establish a sound reputation abroad, particularly among importers. The exporters in Africa made similar efforts and thus helped to consolidate the emerging TOMCs. Their efforts should be positively valued because a dependable, continuous channel is an asset for a less developed economy. However, the issue of reputation also has a negative side. It was natural for well-known businesspeople to speak disparagingly about newcomers. In fact, the established exporters, perhaps unwittingly, made it more difficult for small rivals to acquire a good reputation. This factor should be taken seriously when we read the complaints of African, Indian, and later Lebanese businesspeople about the obstacles they encountered on their way to the position of exporter. These obstacles should be analyzed not merely in legal terms, but as psychological barriers to entry.

Finally, we must point out that forward selling occurred only at the export stage. Everywhere else in the upper section of the TOMC, the moments of sale, delivery, and payment coincided, usually indicated by the term, "cash on delivery" (COD). Accordingly, the need to build up a good reputation (and to retain it) was much greater at the export stage than at the stages farther upstream. It should not surprise us therefore that a contrast arose between the export stage, with its large, well-known companies, and the other stages where small, personal enterprises continued to function satisfactorily.

Glossary

CIF trading [in trans-oceanic trade]: A form of trading in which the exporter is responsible for shipping arrangements and pays the costs of marine insurance and freight.

exporter's reputation [among importers]: A reputation for good performance in executing export contracts. Ideally, a reputation for never defaulting on them.

FOB point: The point of embarkation in the port of shipment. The corresponding FOB value is significant in commercial negotiations and for customs declarations.

forward selling: Here defined as the exporter's practice of entering into an export/sales contract before shipping the contracted products.

trans-oceanic transaction: A transaction between a seller/exporter on one side of an ocean and a buyer/importer on the other. Telecommunications are necessary to bring the transaction about.

Chapter 5

The World Markets
for Trans-Oceanic Crops

In this chapter, we review the concrete, institutional world markets for Tropical Africa's trans-oceanic crops. When comparing them with the abstract world markets of economic theory, two points must be noted. First, while the whole of global supply and demand is supposed to operate in the abstract world market, only a fraction of global supply is sold in the concrete world market. While the transactions and prices in the concrete market are public, the transactions outside the market are called private deals or treaties. (As they are private, contract details such as prices are unknown to rival buyers and sellers.) Second, even if the privately sold volume far exceeds the public volume, the concrete market continues to function satisfactorily as long as it is believed that the price registered in the concrete market equals the hypothetical equilibrium price of the abstract world market.

Below, I distinguish four basic types of market, each associated with a specific group of crops. The first two, the commodity exchanges and the auction markets, are well organized, but this is not the case with the other two. Basic details about the four types of market and crop are given in Table 5.1.

THE TRADE ASSOCIATIONS

We first review the trade associations because wherever a concrete world market was instituted, a trade association stood at its cradle. The early trade associations were private, nonprofit organizations

TABLE 5.1. Four Types of (a) World Market and (b) Trans-Oceanic Crop

Type of Market	Type of Crop	Trade Association	Physical Market	Physical Product
exchange	commodities	strong	yes	absent
auctions	auction crops	strong	yes	present
network	minor crops	weak	no	present
network	perishables	weak	no	absent

Note: A physical market is a place where buyers and sellers meet face to face.

that drew their members from the businesspeople that had a financial interest in the product concerned: producers, exporters, importers, merchants, and manufacturers. Later, most associations became international organizations, often informally by accepting members from many nationalities. As membership is voluntary, the association's strength tends to rest on a broad and balanced membership. This is also necessary to secure confidence and to prevent bias (or the suspicion of bias) in favor of a particular category of members.

A trade association that is responsible for a concrete market sets rules and regulations concerning at least five points. First, the product is demarcated, for instance, in terms of botanical varieties. Second, there are rules about quality (see below). Third, standard contracts and supporting documents are drafted specifying the conditions of a transaction, e.g., a minimum quantity, the latest day for delivery, and the method of payment. Fourth, membership conditions are laid out; these often require a very high financial standing. Fifth, rules concerning arbitration are made. These regulations are designed to reduce the *transaction costs*[1] of buying and selling: the search for a buyer is easy; negotiation is simplified by standard contracts; and nobody needs to worry about enforcement, as all members are trustworthy.[2]

When there is a conflict among members, for instance, about the execution of a contract, the trade association favors settlement by arbitration in accordance with its rules. Members generally abide by the arbiter's ruling because they know that the association will cancel their membership if they do not. Cancellation of membership is a powerful sanction because it entails a serious loss of reputation for

the ousted member. Here, as in Chapter 4, we encounter the significance of having a sound reputation.

Standardization and Grading

Since the late nineteenth century, the trade associations have been
active in the standardization and grading of trans-oceanic crops. The
demarcation of *grades* was not easy because of the natural variation in
shape, size, color, flavor, moisture content, etc., which is inherent to
agricultural produce. For cotton, the length of the fibre became the
main yardstick for grading. For refined sugar, purity and color were the
yardsticks. In the twentieth century, standardization was introduced for
new crops such as cocoa; grading was occasionally revised (e.g., the
maximum number of defective beans permitted in consignments of
cocoa and coffee was lowered); and new yardsticks were defined (e.g.,
for groundnuts (peanuts) in the 1970s after the hazard of aflotoxin was
discovered); and so on. Since all these grades are defined in terms of
objective, measurable characteristics, arbitration on quality is objective
and the ruling of the arbiters is unambiguous.

The number of grades in any standardization arrangement must be
seen as a compromise between the wishes of the buyers and those of
the traders. The latter want to keep the number of grades low because
they know that marketing costs (and for some crops also harvesting
costs) rapidly increase when the number of grades increases. Trans-
oceanic traders tend to be in favor of having few grades.

It may not be amiss to point out that grading has been a private
initiative of the trade associations and that the grades are not based
on national legislation. Neither governments nor international organizations, such as the United Nations and its auxiliary organizations, are empowered to give orders to the trade associations. The
latter have retained their authority in this field by regularly monitoring technical developments and changes in cultivation and by revis-
ing the grading system when necessary.

THE COMMODITY EXCHANGES

Among the concrete world markets, the commodity exchanges for
mineral and agricultural products take pride of place. Table 5.2 lists the

TABLE 5.2. Selected Commodity Exchanges by Crop and Location

cotton	Liverpool, New York, Hong Kong, Tokyo, Osaka
coffee	London, New York, Paris, Le Havre
cocoa	London, Paris, New York, Le Havre
oil seeds	Chicago, Winnipeg, Tokyo, Osaka
rubber	London, New York, Singapore, Tokyo, Kobe
sugar	London, New York, Tokyo

commodities that are relevant for this book. At first sight, it is surprising that there are exchanges for the same commodity in more than one country. If the cocoa exchanges in New York and London both claim to be genuine world markets, why are there two of them, each run by a separate trade association? Is this not wasteful duplication? Apparently, the traders do not think so.[3] There is also a particular type of trade, called arbitrage, that ensures that the prices in the two exchanges remain in step with each other.

The modern exchanges are markets where the product to be sold is not present on the trading floor. This became possible as a result of forward selling, which, in turn, became possible after the installation of the telegraph cables and the standardization of the grades by the trade associations. The commodities lend themselves to *on description trading*, a feature that distinguishes them from, for instance, the auction crops.

When the executives of the trade associations considered the establishment of an exchange, they wanted a market that approximated the world market of economic theory, with full competition and unhindered participation by buyers and sellers all over the world. However, this conflicted with the principle of the physical market where a limited number of buyers and sellers meet face to face and know and trust each other. This conflict was solved by having many brokers among the members. Brokers would take orders from principals scattered across the globe. Thus, while many businesspeople were excluded from the face-to-face trade, nobody was prevented from indirect participation in the market. Of course, the exchanges had to be located in towns with reliable telegraph communications permitting distant principals to contact their brokers without delay. To dispel any doubts about the brokers' integrity, the trade associations actively sought transparency and public scrutiny, which have remained key principles.

Price Quotations

Many newspapers carry a daily column with commodity prices. This is the main form of the price quotations, all of which are derived from the prices for contracts that are concluded on the trading floor. These prices, which are public in the sense that they are cried out, are summarized into a day price. (They can also be summarized for a longer period.) The wide publicity which the price quotations receive is an indication of the authority they have. They are trusted not only by influential businesspeople, but also by governments, although the trade associations, as we saw, operate entirely outside the government sphere. It is especially significant that some international commodity agreements have used the price quotations as thresholds to trigger particular measures of intervention.

Commodity exchanges may be usefully compared to barometers, and the price quotations to the barometric readings. Just as barometers are mainly consulted to anticipate weather changes, so are many commodity exchanges used to provide information about short-run price prospects. This is particularly the case when a terminal market is associated with the exchange.[4] At a terminal market, futures contracts are traded. As these contracts have their own price quotations, there are now several day prices. While the spot price corresponds to current prices, the futures price quotations are prospective prices. Manufacturers and merchants make an active use of the terminal markets to cover or hedge their commercial risks.[5] For the exporters in Africa and elsewhere, the terminal markets are mainly a source of market intelligence. Exporters study the futures price quotations to see what the prices in the coming months are likely to be, and then conclude private deals outside the exchange.

Since all the trans-oceanic crops shown in Table 5.2 are *durable* or storable, traders have to reckon with the present stock situation when they estimate prices three or six months ahead.[6] When current stocks are large, future prices cannot be very different from present ones. But when they are small, a *contango premium* may arise: futures prices may be considerably higher than spot prices.

While in most markets, market intelligence relates to several variables, in the commodity exchanges, it is reduced to one variable: the price.[7] Moreover, as trading is continuous (within opening hours, of course), this new variable is also constantly being registered and consulted. Many businesspeople appreciate the commodity exchanges as a valuable instrument for price discovery. They are willing to pay for the information, whether it is provided directly or through brokers. However, traders are prepared to admit that the instrument is delicate and particularly vulnerable in times of war.[8]

Another vulnerability of the commodity exchanges stems from the freedom that businesspeople, including exporters in Africa, have to trade outside the exchanges.[9] If many businesspeople switch from public transactions on the exchange to private deals, the market becomes a *thin market*, a situation which reduces the representativeness of the price quotations.[10]

THE AUCTION MARKETS

The world markets for a few trans-oceanic crops are characterized by a system of selling by auction. The best known crop in this category is tea. The tea auctions in London date from 1834 and are still important today. Later in the century, international tea auctions were established in the countries of origin, in Calcutta (1861), and in Colombo (1883). A variant of the auction method is the tender market, extensively used in the past for tobacco. Before World War II, important international tobacco markets were located in Amsterdam, Bremen, and Hamburg, where bidding was by written tender instead of by public outcry. Auctions for coffee were started in the 1930s in some countries of origin (see Chapter 12).

To explain why the owners of certain crops prefer this selling method, we must look for any crop characteristics that make auctioning attractive. Typical for *auction crops* is that some significant characteristics are not objective and measurable (in contrast to the commodity grades). Thus, auction crops may be defined as crops for which grading is unsatisfactory as a basis for differentiation.

Traders have been resourceful. Some of them developed sensory standards of taste, smell, visual aspects, etc., to classify the nonobjective characteristics. Thus, a combination of grading and *classifica-*

tion permitted a satisfactory differentiation of the product. This solution is particularly profitable for products such as tea, coffee, and tobacco, where the tastes of consumers may be so special and so refined that the more discriminating among them are willing to pay a substantially higher price for the special qualities they desire in those products.

The use of sensory standards makes inspection before purchase highly important. Intensive training and a long apprenticeship are necessary to become an "expert" or "inspector"—general terms which cover the tea tasters, the coffee liquorers, etc. It is the inspector who must say whether a particular lot or sample meets the sensory standards of a specific class.[11] The inspector's opinion is decisive, for instance, in the bidding at the auctions.

Large organizations pursue interpersonal standardization, which is essential if the organization wants to have a consistent buying policy. Moreover, international harmonization of the sensory standards is desirable, notably for transactions between exporters and importers, because arbitration on quality does not work for auction crops.

Sellers who differentiate their product do so because they expect to earn considerable premiums on the better classes; sellers assume that when all premiums (and discounts for the inferior classes) are taken into account, they will receive more money than they would for the undifferentiated product. As the latter figure is no more than a guess, the gross profitability of differentiation cannot be accurately calculated. There is a further problem when the net profitability is calculated. For this exercise, the sellers must deduct the costs of inspection and the greater logistic complexity of the product flow. Clearly, it is difficult to make a cost-benefit analysis of product differentiation. This is true for the individual seller, for the sellers as a group, and for the government.

I emphasize the problems of calculation for two reasons. First, they explain why the auction method is more easily adopted for the more expensive varieties of a crop. Second, sellers have occasionally introduced auctions for crops they had previously sold as commodities. The opposite has also occurred. Apparently, estimates of profitability, rather than hard figures, have played a crucial role in such changes. To take this into account, I propose the following definition: auction crops are crops for which differentiation on the basis of sensory (or other nonmeasurable) characteristics is profitable in the eyes of influential actors in the TOMC.

THE NETWORK MARKETS FOR MINOR CROPS

We now turn to the *minor crops*, that is, crops for which there is no physical world market. Instead, there is a network of brokers and merchants and hence a concentration of expertise and stocks in one or more ports of destination. As these merchants and brokers are better informed about prices and price prospects than traders else-where, they attract the *network market* to this location. For some crops, they maintain a trade association. A few associations like the FOSFA (Federation of Oil Seeds and Fats Associations) in London publish authoritative weekly price quotations which are based on price estimates provided by brokers and others actively involved in the trade.

The main explanation for the absence of a concrete world market seems to be the low global turnover for these crops, which leads to intermittent (rather than continuous) trading and makes the market too thin. These crops have often been called the minor crops, in contrast to the major crops traded on the exchanges and at the auctions. In the absence of a physical market, all trade in the minor crops is conducted by means of private deals. The unifying effect of competition is therefore weaker than on the commodity exchanges.

THE NETWORK MARKETS FOR PERISHABLE CROPS

Perishable crops such as bananas are here defined as crops that require special refrigerated ships to transport them. The trade is characterized by the overruling necessity of having the product reach the final consumer as rapidly as possible. It follows that there are no merchants for these crops. The network market consists of brokers only. They are engaged in a price discovery process. It is further relevant that the trans-oceanic trade in perishables such as bananas is not conducted consignment by consignment, but on the basis of periodic contracts stipulating regular shipments.

There are no current price quotations that correspond to the theoreti-cal world price. Any price data that are published are prepared by customs departments on the basis of export documents which refer to a date long after the one on which the contract was signed.

COMMENTS ON THE NEW CLASSIFICATION

How are the four categories of world market, identified above, related to each other? We will look first at the analytical relationship. The following questions help to classify the crops unambiguously. First, is the crop perishable at the port of shipment? If it is, it belongs to the perishable crops. If it is not, the crop is durable. Second, for durable crops, is inspection prior to purchase necessary? If it is, the crop must be classified as an auction crop. If it is not, it belongs to the *gradable* crops. Third, for the gradable crops, does a commodity exchange exist for this crop? If it does, the crop is a commodity. The minor crops remain as a residual category.

Next we examine the historical relationship. The auctions and the network markets for minor crops have the longest history. They were found in several European *entrepôt ports* in the sixteenth century, perhaps even earlier. The exchanges are much younger because they had to wait until CIF trading and selling on description became possible. Some crops which we now call commodities seem, in the days of consignment sales, to have been sold at auction markets. The auction method was popular and helped to clear newly arrived consignments quickly. The only difference between what we now call commodities and auction crops may have been that the prior inspection was more thorough for the latter.

When CIF trading grew in significance, some crops qualified for a grading system and became commodities (in the narrow sense used in this book). The auctions for those crops declined, but for the crops without a satisfactory grading system, exporters, in Africa, for instance, could not switch to sale by cable (because the importer wanted to inspect the consignment before buying it) and the auction method remained viable.[12] Presumably, the trade in undifferentiated (or bulk) commodities grew more rapidly after 1870 than the trade in the highly differentiated auction crops. The resurgence of the auction method in the 1930s (see Chapter 12) may signify a reversal of this trend.

There is little to be mentioned about the world markets for perishable crops. As trans-oceanic trade in these products started late, there was no scope for concrete markets before World War I. The network markets that later developed were feeble. The absence of merchants is one factor. However, there was also a change in the general economic climate which was less favorable for institutional solutions.

How does my classification relate to other classifications? First, the distinction between perishable and durable crops is well known, but it is mainly used to highlight the perishables. My purpose is rather to stress that the "storability" of durable crops is a positive feature: it made these crops suitable for trans-oceanic transport long before the perishables could be exported. Second, the recently introduced adjective "nontraditional" seems unsatisfactory to me because it is so vague. Third, the contrast between food and industrial crops, while useful from a demand perspective, is not commercially relevant. In the same way, the beverages category is not helpful. Fourth, the difference between annual and tree crops, while important for supply analysis, hardly matters for the organization of the market. In Table 5.3, specific products are listed for each crop category.

It may seem strange that coffee and tobacco appear in two columns. It is less strange when we remember that mainly the expensive varieties (arabica coffee and Virginia tobacco) are sold at auctions. Although groundnuts and palm kernels are important African oil seeds, they have never been traded on an exchange. Nevertheless, they are here classified as commodities because of the easy substitu-

TABLE 5.3. Selected Trans-Oceanic Crops, Newly Classified

Commodities (1)	Auction Crops (2)	Minor Crops (3)	Perishable Crops (4)
cotton	tea	sisal	bananas
coffee	tobacco	tobacco	pineapples
cocoa	coffee	cashew nuts	
oil seeds		cloves	
rubber		vanilla	
sugar		ginger	

tion of one type of oil seed for another and the fact that some oil seeds such as soybeans, sunflower seed, and rape seed have long been traded on exchanges in North America.

A CORRESPONDING CLASSIFICATION FOR TOMCs

For each of the columns in Table 5.3, there is a characteristic type of TOMC. These four types structure the discussion in Chapters 7 through 13. Here a few preliminary points about (a) the commodity TOMCs and (b) the perishable crop TOMCs are made because in many ways, these two types of TOMC form a contrast.

The special features of the commodity TOMCs are largely determined by the commodity exchanges. Elsewhere (Van der Laan, 1993a), I have argued that they divide the commodity TOMCs into two truncated, essentially independent parts. The exporters benefit from this division because they can operate effectively, even if they know little or nothing about the lower half of the TOMC (between the exchange and the consumer). A similar conclusion applies to importers. I therefore called the commodities *half-channel crops*. The exchanges may be compared to a smoothly operating global turntable. Any half-channel in an upper section may be quickly connected to any half-channel in a lower section. (This observation applies to both countries and individual exporters.) One advantage for manufacturers in the lower section should be specifically mentioned. These manufacturers no longer have to worry about their supplies, not even when there is a hostile rival who would like to cut them off from long-established sources of supply. The commodity TOMCs are therefore conventional marketing channels as defined in Chapter 2. Put differently, vertical, ocean-straddling integration, for instance, when manufacturers established plantations to grow their own supplies, have been rare.[13]

The commodities are known for their high *marketability*. Marketability is said to be high when it is easy for a seller to find a buyer at any time, as is the case on the commodity exchanges. This is a great advantage for the exporters.[14] True, they may be obliged to accept a lower price, but in the eyes of most exporters this situation is better than that in the other markets where intermittent trading is the rule and where they might not be able to find a buyer at all, however low the price they are willing to accept. It is logical for exporters to have an ambivalent attitude toward the commodity exchange. They detest the fluctuating prices, but treasure the continuous trading. Of course, these features are inseparable.

We will now turn to the TOMCs for perishable crops. Typical for these TOMCs is the elaborate coordination between producers, exporters, shipowners, and importer-distributors. Each of them has to study the entire channel. That is why I have called the perishable crops *entire-channel crops*. The ocean-straddling coordination often leads to a vertical marketing system. These TOMCs tend to be less competitive and less transparent. Moreover, the absence of stocks (apart from the stocks in the pipeline) leads to a different process of price formation. In these TOMCs, MCA is more helpful than market theory. Market failure, that is, the failure of a free market to operate as postulated in the theoretical models, is also quite common. With regard to the other TOMCs (for minor and auction crops), it suffices to say here that they resemble the commodity TOMCs in some ways and the perishable crop TOMCs in others.

From the exporter's perspective, the commodities are *easy-to-export* and the auction, minor, and perishable crops are *difficult-to-export*. However, this should not create the impression that it is easy to make profits in the commodity TOMCs. On the contrary, profit margins are low because competition is pervasive and actors have to be skilled and alert if they want to survive. This is one of the reasons why I devote four chapters (7 through 10) to the discussion of the commodity TOMCs. (The other reasons are that they have dominated both Africa's exports and the literature on them.) Only three chapters are needed to examine the difficult-to-export crops, partly because they can be contrasted with the commodities and partly because two categories (the auction crops and the perishable crops) appeared relatively late on the scene, around 1930.

Glossary

auction crops: Crops for which product differentiation on the basis of sensory (or other nonmeasurable) characteristics is profitable in the eyes of influential actors in the TOMC.

classification: Here defined as the supplier's activity of differentiating his product on the basis of sensory (or other nonmeasurable) characteristics.

contango premium: The (irregularly occurring) margin of futures prices over spot prices.

difficult-to-export crop: A crop whose marketability in international trade is low.

durable: Here defined as the equivalent of storable and the opposite of perishable. This reflects the view of growers and traders rather than those of consumers.

easy-to-export crop: A crop whose marketability in international trade is high.

entire-channel crops: Trans-oceanic crops for which the world market is poorly organized. An exporter is therefore obliged to collect a great deal of information about the lower section of the TOMC in order to operate successfully.

entrepôt ports: Ports where stocks of products, in particular trans-oceanic products, are concentrated. These ports have facilities for transshipment and for duty-free storage (in bonded warehouses) prior to reexport.

gradable crop: A crop for which a satisfactory grading system has been developed.

grades: Divisions or categories of a raw material; each grade is demarcated by, and described in terms of, measurable standards of quality.

half-channel crops: Trans-oceanic crops for which the world market is highly effective in bringing supply and demand together. Exporters therefore do not need to study the lower section of the TOMC. Similarly, importers do not need to study the upper sections.

marketability: A measure for the ease of marketing. A crop is highly marketable when the seller has to make hardly any efforts (and in any case, no promotional efforts) to find a buyer.

minor crops: Here defined as a residual category of durable and gradable trans-oceanic crops. The world markets for these crops are network markets.

network market: Here defined as an international market whose operation primarily depends on a network of brokers and merchants who possess superior market intelligence with regard to the product concerned. Typical for a network market is the absence of a public marketplace where buyers and sellers meet face to face. Consequently, most network markets are less well organized and less efficient than the commodity exchanges and the auction markets. A network market tends to be centered in a port of destination.

on description trading: Selling and buying on the basis of internationally recognized grades or another mutually acceptable description. This form of trading is convenient when buyers and sellers are far apart geographically. It is the opposite of trade on the basis of prior inspection of (a sample of) the goods.

perishable crops: Here defined as crops that are perishable at the point of shipment and therefore require refrigerated ships to transport them across the ocean.

thin market: A concrete market in which the number of transactions is so low as to undermine the authority of the price quotations.

transaction costs: A group of transaction-related costs such as the costs of buyer search, negotiation, and contract enforcement.

PART II:
THE TOMCs OF TROPICAL
AFRICA'S EXPORT CROPS

Chapter 6

Three Upper-Section Patterns— Commercial Practices

In Chapters 6 through 13, I describe and analyze the TOMCs for Tropical Africa's export crops. In doing so, I draw on empirical material from many African countries over a long period. In this first chapter, we examine some features which these TOMCs have in common. Most of these features depend on the number of successive actors in the upper section of the TOMC. Accordingly, we distinguish three patterns: Pattern 1 with one actor, Pattern 2 with two actors, and Pattern 3 with three (or more) successive actors. In this chapter, we discuss them in turn.

Among the special features of Patterns 2 and 3 there are various commercial practices which are sharply distinguished here from the commercial policies discussed in later chapters. Most economists are primarily interested in policies and may be tempted to skip this chapter. It is my view, however, that the policies cannot be properly studied unless one knows and understands the practices.

THREE PATTERNS

Figure 6.1 shows three basic upper-section patterns. Figure 6.1a illustrates the *one-actor or grower-exporter pattern*, Figure 6.1b illustrates the *two-actor pattern* (with growers and exporters), and Figure 6.1c illustrates the *three-actor pattern* (with growers, intermediaries, and exporters). Our discussion begins with the one-actor pattern and ends with the three-actor pattern. This is no reflection on the economic significance of these patterns. In fact, in terms of export volume and value, the three-actor pattern has been predominant in Tropical Africa. Before taking each pattern separately, it is helpful to emphasize two contrasts. First, Pattern 1 is contrasted with the other two patterns and then Pattern 2 with Pattern 3.

The first contrast is clear and generally understood. The *grower-exporter* tends to be large, sophisticated, and entrepreneurial in outlook. He directly participates in international trade and faces the commercial risks involved. External marketing is therefore important for him. Since the grower-exporter combines cultivation and exporting he is in a position to achieve a high degree of coordina-

FIGURE 6.1. The Basic Upper-Section Patterns in the TOMCs

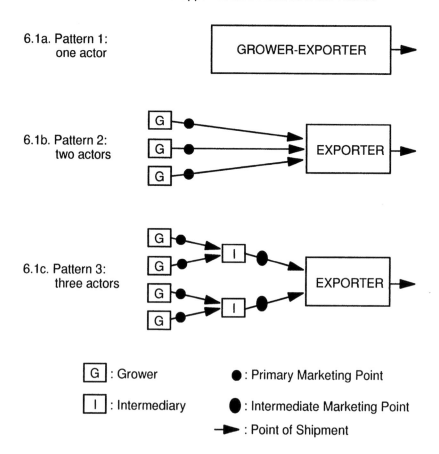

tion of production and marketing activities. The upper section is under unified management.

By contrast, the growers of the other two patterns tend to be small in terms of volume handled. The average exporter buys from many growers and the convergence factor is high. (The factor suggested by the diagram is too small.) In terms of MCA, there are channel systems and channel leaders. The strength of the channel leader is greater when the growers are numerous and small. Besides external marketing, there is internal marketing, that is, marketing upstream of the port of shipment.

We now turn to the contrasts between Patterns 2 and 3. The main difference is that there is a middleman or intermediary in Pattern 3. As a result, trade takes place at two levels: at the *primary marketing points* (PMP) and at the *intermediate marketing points* (IMP). Convergence may be observed at two stages. In the early colonial period, exporters referred to Pattern 2 as that of *direct trade*, which they preferred to *indirect trade* represented by Pattern 3. The term "indirect trade" was also used for situations with trade taking place at three or more levels.

In general, the exporter in Pattern 2 is a more powerful channel leader than the one in Pattern 3, partly because he faces relatively small sellers and partly because competition among exporters tends to be absent in this pattern. (Each exporter operates in his own, often small, exclusive area.) It follows from the MCA perspective that exporters want to preserve direct trade and oppose the emergence of middlemen. However, the historical record shows that the number of middlemen increased, notably in the period after World War I. The general interpretation has been that direct trade was an unrealistic strategy and gave way to indirect trade, as the channel systems expanded, with larger volumes handled and greater distances covered by inland transport.[1] But this interpretation misses an important point, which we must now analyze.

The exporter's reluctance to accept middlemen easily weakened, if only commercial considerations were relevant. Savings on costs were then enough to sway his opinion. When there were also technical considerations, the exporter tended to defend direct trade. Technical points were either a matter of improved cultivation or processing, as with cotton and coffee. Such exporters invested in processing, and engaged in the provision of farm inputs, extension, and supervision. (In the cotton sector, all of these points have been relevant.) Since World War I, the typical exporter of Pattern 2 has therefore been a *processor-exporter*, while the "pure" *trader-exporter* was characteristic for Pattern 3. These terms will therefore be used here.

In the literature, the processor- and trader-exporters are often taken together. However, this ignores the fact that the economic interests of the former are stronger than those of the latter (assuming the same turnover for the two actors). As a result, processor-exporters do not easily withdraw, which is an advantage for the continuity of the TOMC. At the same time, these interests prompt processor-exporters to retain or strengthen their channel control. We must therefore expect various policies and instruments to keep the growers captive (see Chapter 8).

A trader-exporter is, as we saw, a weaker channel leader. He has to cope with two problems. First, the middleman forms a complication in the channel and often weakens his control. Second, most of the time, trader-exporters compete with each other, which tends to show up any weak spots in channel control. As a result, the issue of competition (and the attempts to curb it) have been more important in the literature on Pattern 3. This is elaborated in Chapter 9.

THE ONE-ACTOR PATTERN: GROWER-EXPORTER

Plantations have been the most conspicuous form of the grower-exporter pattern in Africa. Since plantations are large cultivation units, it is worthwhile for the owners to have their own exporting departments. This has been the case for sugar (on Mauritius and Réunion and later in Mozambique), sisal (in East Africa), cocoa (on Fernando Po, São Tomé, and in German Kamerun), oil palm (in the Belgian Congo), tea (in East Africa), rubber (in Liberia and Nigeria), and bananas (in Nigeria). Nearly all of these plantations were corporate enterprises, owned by large parent companies in Europe and North America. This type of colonial investment came to an end in 1929 (Lewis, 1978b, p. 44). It was some 30 years later that interest in plantations revived, with rubber as the principal export crop.[2]

Another form of the grower-exporter pattern, which existed mainly in the colonial period, involved agricultural units which were smaller than the corporate plantations. Most of these units were family-owned farms or estates, on which the families lived. Typical were the settler farmers in British East Africa, but there have been similar entrepreneurs in the French, Italian, Belgian, and Portuguese colonies. Measured in terms of production, settler farmers were certainly not large entrepreneurs, but many of them had the sophistication to understand the intricacies of external marketing. Those who were unable to undertake their own marketing, employed commercial agents who took care of shipping, finance, and selling. From the account that Stahl (1951) gives for the coffee and sisal growers in East Africa in the interwar period, it is clear that principals and agents sometimes disagreed and that the coordination of the commercial activities was far from perfect. Nevertheless, these entrepreneurs should be classified as grower-exporters because they exported on their own account.

The grower-exporter pattern has been relatively rare in Africa. This conclusion corresponds with the generally held view that, in a comparative global perspective, the number of plantations in Africa has been small. In this connection, it is significant that in two sectors, plantations were almost completely replaced by small farms. In the cocoa sector, this happened before World War I and in the robusta coffee sector after World War II.

It would appear that conditions for export plantations have generally been unfavorable in Africa. The explanation most often advanced is the scarcity of labor and, consequently, the high labor costs.[3] This explanation should not go unchallenged. I believe that MCA offers an additional hypothesis which may be called the risk-separating explanation. At least two types of risks are to be considered: the production risks of the grower and the commercial risks of the exporter. In the grower-exporter pattern, both types of risk are borne by a sole actor, whereas in the other patterns the risks are separated. Assuming that risks were perceived to be higher in Africa than elsewhere, the natural reluctance of investors to cumulate risks would explain the tendency for the other patterns to develop.

THE TWO-ACTOR PATTERN

It is typical of the two-actor pattern that growers and exporters are separate actors and that they need each other. Without the growers, the exporters would have no crops to trade in; without the exporters, the growers would be unable to dispose of their product, because their volumes are too small and they lack the skills necessary to engage in trans-oceanic trade. While this observation suggests a symbiosis between growers and exporters, conflicts of interests often arise.

All the transactions between growers and exporters constitute a set of internal marketing practices. Nearly all these practices have their origin in the formative period, the period from 1880 to World War I, a time when direct trade (exemplifying the two-actor pattern) was still important. In this period, an exporter was a channel leader, for both structural and incidental reasons. It is a structural principle that in the upper section of a TOMC the downstream actor is always larger and better informed than the next adjacent upstream actor. Thus, the exporters during this period were larger because the channels converged, and the exporters were better informed because they knew the rules of trans-oceanic trade. The incidental factor was the European origins of many exporters. This gave them prestige which fostered acceptance of their status by the growers.

Commercial Practices Introduced by the Exporters

Table 6.1 presents information on the commercial practices in internal marketing. Column 1, the only column we consider for the time being, shows the dominant role of the exporters. Exporters introduced and imposed measurement by weight, a practice foreign to the growers, who were used to measurement by volume. Exporters also provided the scales and balances and did the weighing. It is significant that the growers accepted the practice of weighing by the buyers, even if they had misgivings about the outcome of the weighing process.[4] In much the same way the exporters translated the quality standards of international trade in their dealings with the growers: the growers submitted their produce to a quality examination and learned that some parcels could be rejected. This may have seemed arbitrary to the growers, while the exporters would view it as an obvious way of

TABLE 6.1. Commercial Practices in Internal Marketing

	Direct trade at PMP (1)	Indirect trade at PMP (2)	Indirect trade at IMP (3)
1. Weight measurement imposed by	exporter	intermediary	exporter
2. Weight measured by	exporter	intermediary	exporter
3. Quality standards set by	exporter	intermediary	exporter
4. Quality examined by	exporter	intermediary	exporter
5. Time of examination	immediately	immediately	often later
6. Price translated by	exporter	intermediary	exporter
7. Price paid	set by exporter	negotiated	sometimes negotiated
8. Purchase location decided by	exporter	intermediary	exporter
9. Timing of payment	COD	COD	advances
10. Receipts/documents	none	none	standard
11. Parcel/load size decided by	grower	grower	intermediary
12. Packing/container decided by	grower	grower	exporter
13. Time of delivery decided by	grower	grower	intermediary
14. Negotiation, discretion of buyer	none	some	much
15. Personal relationships	none	some	much
16. Cheating	much	some	little
17. Redress in case of cheating	none	some	much
18. Credit and trust	none	some	much

PMP = primary marketing point
IMP = intermediate marketing point
COD = cash on delivery

doing business. The price paid to the growers was derived from the world price—a price known only to the exporters.

In three other respects, independent of international commercial practices, the exporters could have their own way. First, they were free to select the locations, the primary marketing points (PMPs), where they took delivery of the growers' crops. Thus, a pattern of what I have called *stationary buyer, mobile seller* developed.[5] Second, exporters could choose between negotiated or set prices when buying. In fact, the set price became the universal pattern in Africa. In view of the large number of growers per exporter[6] the average exporter was loath to negotiate with each grower separately. Exporters preferred a uniform buying price on an impersonal, take-it-or-leave-it basis. Third, exporters were free to introduce a standard practice for payments. In Africa, this was generally prompt payment or cash on delivery (COD).

It may seem that little room for choice was left for the grower. This is true, but the points on which growers retained freedom deserve mention. First, the growers were free to chose the moment of sale. They came to the PMP when it suited them. (Since the growers offered their products as soon as possible after the harvest, the exporters could predict when the buying season would start and make the necessary preparations.) Second, the growers were free with regard to the quantity—the *parcel*—they offered for sale. The growers, in any case, lacked equipment to measure by weight before their trip to the PMP. (In keeping with West African usage, I use the term "parcel" to refer to the quantities offered by the growers at the PMPs.) Third, the growers could use any containers they had for transporting their parcels. The containers would then be emptied, usually on the ground, where the product was inspected prior to purchase.

Cultural barriers, notably the language, prevented the relationships between the exporters (mainly European) and the growers (mainly African) from becoming personal. The language barrier generally ruled out the possibility of exporters' negotiating prices with growers. The exporters saved valuable time, but lost the chance to develop personal contacts. Exporters rarely knew the names of the growers and did not issue receipts recording these names. The COD practice reinforced the impersonal relationship. It meant that the exporters had no redress if, after having paid, they later discovered that the quality of the parcel was substandard. Thus, an immediate quality examination assumed critical importance for exporters. On the whole, transactions remained incidental and anonymous—conditions which easily engendered cheating on either side[7] and an atmosphere of distrust. It made direct trade unstable.

The practice of *COD at the PMP* remains a widespread feature in Africa. Nearly everywhere, it has become a normative rule which buyers cannot violate without antagonizing the growers.[8]

The Time Factor

Having examined the practices at the PMPs, let us now consider the *primary marketing moment*, the moment at which the transaction takes place. Prior to this moment, there is a postharvest or on-farm interval. It is further useful to distinguish growers who sell their whole harvest in one parcel from those that bring a number of parcels on successive occasions. As the growers are free to choose the size of the parcel, they are the ones who decide whether the quantity harvested is large enough to justify a trip to the PMP. A small harvest, a short harvesting season, and a long distance to the PMP are all conditions that favor only one parcel and one trip, and thus only one marketing moment for the grower. (The remainder of this section presupposes a one-grower-one-parcel situation, but the reader should be aware of the alternatives.) The postharvest interval varies among crops and for the most part, depends on the need for on-farm preparation (cleaning, drying, sorting, processing, measuring, bagging, etc.[9]). No less important is the growers' wish to dispose of their crop quickly or not. In the long-standing experience of exporters, most growers want to sell as soon as possible.

Theoretically, the primary marketing moment marks the beginning of the *exporter's interval* for a parcel. However, since the growers deliver many parcels, an exporter has to work with an average interval. During this interval, he prepares the product for export by standardizing quality and weight to satisfy international standards. In some cases, an exporter must also process the product, which makes the interval much longer. The standardized product is transported in lorry or railway wagon loads. At this point, the term "load" or "consignment" is appropriate, as opposed to "parcel."

The internal marketing of most crops is seasonal in character. If an exporter wants to be equipped to handle the volumes of the peak months, he will inevitably have idle capacity during the slack months. This is true for his assets such as buildings and means of transport, but also for capital and staff. It can therefore make sense for an exporter to employ a small permanent staff, seasonally supplemented by casual laborers. In the same way, it may be cheaper for him to borrow some money during the peak months rather than to finance everything from his own resources. We will come back to this point, but here it must be stressed that the seasonal element complicates an exporter's accounting system and makes the calculation of a reasonable trading margin difficult.

The Exporter's Trading Margin

An exporter's *gross trading margin* is defined as the difference between his selling and his buying price. Normally, the margin is calculated after the buying and selling transactions have taken place: this is the *ex post* margin. However, because of the commercial practices just discussed, we must also consider the *ex ante* margin. The pervasive practice of COD at the PMP compels an exporter to pay before he knows his selling price. (The problem is exacerbated by the long internal marketing interval which is typical for the TOMCs.) However, since exporters are price setters, they have to fix their buying price before they start buying!

There are three steps in an exporter's price-setting procedure: He first makes an estimate—no more than a reasoned guess—of the price he expects to obtain when he is ready to sell. He then subtracts the *ex ante* trading margin he wants to earn, and arrives at the buying price, which he then sets. This pricing policy is satisfactory if the exporter's estimate turns out to be correct several weeks later. The realized price then equals the estimated price, and the *ex post* trading margin equals the *ex ante* margin. Problems arise when the world price falls during the interval. An exporter's *ex post* trading margin is then eroded or eclipsed. Clearly, the *ex post* trading margin is largely unpredictable. Given the uncertainty about world prices (see Chapter 7), the exporter is tempted to set a low price in order to have what I call a *contingency margin* as well as a reasonable trading margin.

Outgrowers

Before discussing the three-actor pattern, the case of the outgrowers should be briefly reviewed. The term *outgrower* is relatively recent and refers to small-scale farmers who have a special relation with a planta- tion company.[10] The company operates a nucleus plantation and pro- cessing facilities. As the production of its own plantation is lower than its processing capacity, the company is willing to buy additional sup- plies of unprocessed product from the outgrowers. In our terminology, the outgrowers' case can be described as a two-actor pattern grafted onto a one-actor pattern. This formula has worked well in several countries. For rubber, interesting information on the Liberian experi- ence in the 1970s was collected by Carlsson (1983).

THE THREE-ACTOR PATTERN

The increased complexity of the three-actor pattern (and the three-or-more-actor pattern) over the two-actor pattern derives first of all from the presence of an additional actor, the intermediary. Intermediaries have their own interests which they defend in two directions: upstream at the PMPs and downstream at the IMPs, where they meet the exporters. These complications have made for an extensive and polemical literature on the intermediaries. Coordination of the channel system is more difficult. In the absence of direct contact with the exporters, the growers tend to have (or to expect) more freedom. It is more difficult for the channel leaders to gain or retain control of the channel system. The strategies channel leaders design are more circuitous and harder to investigate (see Chapter 9).

The term "intermediary" applies to all actors between the growers and the exporters.[11] It includes the middlemen of the past, that is, of the period before World War II, and the licensed buyers who collaborated with the public exporters after World War II. Since licensed buyers only dealt with a single public exporter, they were more restricted in their economic freedom than the earlier middlemen who could deal with many private exporters. In the present chapter, the emphasis is on the interwar period and on intermediaries who collaborated with private exporters.

In many areas of Africa, the three-actor pattern was the natural successor of the two-actor pattern. As more growers, particularly in remote areas, began to cultivate export crops, exporters saw the share of direct trade decline and that of indirect trade increase. Put differently, as the number of PMPs in the country rose, the number manned by intermediaries increased more than proportionally. As the individual intermediaries were too small and weak to impose their own practices, they generally followed those introduced earlier by the exporters. We will first discuss the practices at the PMPs and then those at the IMPs.

Commercial Practices at the PMPs

It is now time to consider Columns 2 and 3 of Table 6.1, which deal with indirect trade. Column 2 shows that the new intermediaries largely copied the practices of the exporters. If an exporter distinguished only one grade, the intermediary's work was simple.[12] If there were two or more grades, an intermediary's work was more difficult because he then had to keep these grades separate in the assembling or bulking process at the PMP. In differentiating the produce and thus the price, an intermediary had to know a downstream price and to translate it into an upstream one. Having up-to-date information about the exporter's buying price and how it was set was therefore important to an intermediary.

The main change brought by the intermediaries was the personalization of relationships at the PMPs. Insofar as they were Africans, it has often been supposed that lower cultural barriers facilitated this change. However, overseas intermediaries, such as the Lebanese in West Africa and the Indians in East Africa, also had a reputation for having better social relations with the growers than the early European exporters. I attribute the change to the willingness of many intermediaries to negotiate prices.[13] The intermediaries thus modified the take-it-or-leave-it approach of the early exporters. As they got to know the growers better, relations of trust developed, which in turn favored the granting of credit, largely of the preharvest type. No doubt, getting *preharvest credit* was appreciated by the growers, but it was also the first step toward a situation of generalized rural indebtedness.[14]

If prices were negotiated, the concept of a uniform buying price at the PMPs was obviously not tenable. It was, at best, the average of the prices agreed upon by all the different intermediaries with the various growers. Obviously, such an average is hard to measure or even estimate. This may well explain why before World War II, there was great disagreement among observers about this price and the level of growers' incomes. Pessimists held that the intermediaries were very clever and used negotiation to reduce prices. Optimists held that competition, if present, protected the growers. In any case, negotiation led to a differentiation among the growers: the influential and the experienced got a better deal than the diffident and the newcomers.[15]

Commercial Practices at the IMPs

Intermediaries and exporters regularly met at the IMPs. Column 3 of Table 6.1 summarizes the information about the practices that were observed at the IMPs. Many of those practices were imposed by the exporter. Not only did the exporters operate on a larger scale than the average intermediary, but they also possessed more skills, for instance with regard to exporting and shipping. In general, intermediaries were weak because their knowledge of world prices was small. Nevertheless, an exporter was not an unrestrained channel leader who could dictate his own terms to the intermediary; they needed each other.

An exporter needed the intermediaries primarily to alleviate his own work during the peak months of the buying season. If an exporter had been obliged to examine every incoming load carefully, he would have needed far more personnel. Examination could be simplified and partly postponed if the exporter bought from trusted and experienced intermediaries. A new unit of trade emerged: the sown-up bag of produce, filled to standard weight. Exporters saved considerable time working with such bags, as long as the intermediaries had done their job well. Final grading (i.e., before export) became simply the verification of the grading performed earlier by the intermediary. Fumigation before shipment became a mere routine job, as long as the intermediaries had been thorough in disinfecting their stores. Smooth collaboration was a valuable time-saver.

Trust was an indication of good personal relationships, at least with those intermediaries who performed well. (Transactions at the IMPs were accompanied by documents which reinforced the personal element.) Two consequences should be noted. First, trust favored credit. There was indeed a large flow of credit from the exporters to their trusted intermediaries during the interwar period. Second, a certain readiness to negotiate gradually developed. Of course, negotiation was a slippery path for the exporters because a higher buying price reduced their own trading margin. It was a zero-sum game: a higher trading margin for an intermediary led to a lower margin for the exporter, and vice versa. However, negotiation could not be avoided when the intermediary was experienced.

The Time Factor

In the three-actor pattern, there are three intervals instead of the two intervals found in the two-actor pattern. If we assume that the grower's (post-harvest) interval is practically the same as in the two-actor pattern, we may confine our attention to the *intermediary's interval* (which is new) and the modified exporter's interval. The intermediary's interval is relatively short, partly because the geographic distances to be bridged are short and partly because the intermediaries, like the growers, want to sell as soon as possible.

The literature prefers to think in terms of an *internal marketing interval*, which combines the intervals of intermediary and exporter. This internal marketing interval is a valuable concept, but it is hard to quantify.[16] Even insiders are reluctant to estimate it because conditions in Africa are so diverse. To readers who insist on a rough idea, I suggest that on average, the whole interval lasts six weeks, with the intermediary's interval being one week. (Time-wise, the exporter plays a larger role in internal marketing than the intermediary—some five times as large.) Both intermediaries and exporters try to shorten the internal marketing interval. As the primary marketing moment cannot be shifted (it is fixed by the weather, the time of harvest, and the growers' inclination to sell quickly), all attempts of the actors are directed at advancing the intermediate marketing moment and the moment of shipment (see Chapter 10).

The exporters' operations have always been better documented than the intermediaries.' Thus, the marketing costs, including transport costs, could be verified in the exporters' accounts, while those of the intermediaries were hard to ascertain. The same holds for the marketing margin. In the case of the intermediaries, the margin was elusive because, as we saw, the average buying price of the intermediaries was hard to calculate. This lack of reliable data has given rise to wild speculations about the profits made by intermediaries. It also casts doubts on any figures for the *ex post* trading margin of the intermediaries.

The seasonal problems of the three-actor pattern are also more complex. For instance, the intermediaries tried to shift the burden of seasonal finance to the exporter. On the other hand, many intermediaries owned lorries and faced the problem of finding a use for them in the slack months. Exporters could turn to rail transport, which they could reserve and use when needed. In doing so, they shifted a seasonal burden to an actor outside the channel, the public railways, which were left to solve the problem of overcapacity in the *off-season*. How the costs of seasonally idle resources are shifted to other actors (in or outside the marketing channel) is a fascinating subject from the macroeconomic point of view and hence important for the government.

To be complete, we must say something about the intermediaries in the patterns with more than three actors. In a four-actor pattern, there are two successive intermediaries: a smaller one upstream and a larger one downstream. There is also an additional interval and trading margin. In the pre-1940 literature, it is often asserted that there were too many intermediaries or middlemen in Africa. The accumulation of their trading margins, it is suggested, lowered the price received by the growers. (These remarks refer to successive intermediaries and not to parallel ones.) Later, several tentative explanations for a long channel and many successive intermediaries were advanced. First, as distances in Africa are long and growers dispersed; it would be easier for two independent intermediaries to bridge this distance than for one integrated enterprise to do so. Second, ethnic variety and cultural differences would favor a long channel. Third, the dearth of capital made it more difficult to estab-lish a large enterprise (downstream) than a small one (upstream). It follows that, when traders who started on a small scale in a village, became successful enough to move to a town, this has been consid-ered as natural progress and as a sign of capital accumulation. Whatever the explanation, the fact remains that in many channels, the number of successive intermediaries was unknown and so was the accumulated trading margin.

MACROECONOMIC ASPECTS

We end this chapter by briefly reviewing the areas of overlap with research done by economists and geographers. The contribu-tion of macroeconomists normally begins with the foreign trade statistics, which provide useful information on the annual tonnage and value of each export crop. Let us consider the volume figures first. For cocoa (and many other crops), the tonnage harvested will equal the tonnage exported, at least if certain conditions are met (no internal consumption and no carryover stocks from the previous year). While the aggregate figure may be reliable, it reveals nothing about the individual actors or channel systems. Also, the value figures help to estimate the amounts paid to the growers either by the exporters (Pattern 2) or by the intermediaries (Pattern 3). The figure for the growers' aggregate income can only be a rough estimate because, as we have seen, the trading margins of the exporters and intermediaries are difficult to determine.

GEOGRAPHICAL ASPECTS

Geographers have made a significant contribution to our understanding of the three-actor pattern. In Figure 6.1c, the channels converge, first at the PMPs, then at the IMPs, and finally in the port of shipment where the exporter operates. Convergence is also the element highlighted by geographers. The product flow starts in the form of many small parcels originating with many growers. Later, these are combined (bulked) into larger quantities, and finally, the entire product flow arrives in the port of shipment. In Chapter 2, the metaphor of the river drainage system was introduced.[17] It emphasizes, on the one hand, the fragmentation in the production areas (caused by the low productivity of small-scale agriculture and the low degree of specialization in export crops) and on the other hand, the concentration of stocks in the port of shipment. The contrast between fragmentation upstream and concentration downstream is a recurring theme in the geographical literature.

The model of the geographers is almost identical with the one sketched in this chapter. The agreement is greatest when there is (1) only one port of shipment; (2) one export crop; and (3) one exporter in the country. Even then geographers have their special interests. Thus, while MCA focuses on the transactions and the points where these take place, geographers look at transport. Geographers distinguish three transport stages in internal marketing: from farm to PMP, from PMP to IMP, and from IMP to the port of shipment. Accordingly, I propose to call the area between the PMPs and the IMPs the *Outer Zone*, and the area between the IMPs and the port of shipment the *Inner Zone*. In the Inner Zone, the means of transport (trains, large lorries, and river vessels) have a high load capacity. In the Outer Zone, smaller lorries predominate, while between the farm gate and the PMP, the simplest form of transport is to be found, normally organized by the farmers themselves. Their main contemporary means of transport are pick-up lorries. In the past, they used bicycles, pack animals, canoes, or carried loads on their heads in traditional fashion.

Trains, river vessels, and lorries are tied to well-defined routes. This makes the metaphor of the river drainage system so telling. As the river bed determines the way the water will drain off, so do navigable rivers, railways, and roads determine the way the produce will "drain off" to the port of shipment.[18] This has caused great rigidity in the pattern of convergence.

A notable activity at the IMPs is the transfer of the product from one mode of transport to another; for instance, from lorries to trains. But prior to the transfer at all the IMPs (and to a lesser extent at the PMPs) the product is temporarily stored. The product flow is thus characterized by an alternation between transport and storage which has a bearing on the rate of evacuation (see Chapter 10). The river drainage metaphor does not capture this stop-and-go element, but it does illustrate the seasonal factor well: the river bed is often dry for several months because there is no product at all, then the product begins to flow until it reaches full spate; it finally declines and dwindles to nothing, and so on, in an annual cycle.[19]

The condition of having a single port of shipment is met in most African countries. Many ports were built or greatly improved at the beginning of the colonial period. It was part of the colonial design that the whole territory was seen as the hinterland of this port. However, while a river drainage system remains a constant, a hinterland does not; it can be extended farther inland by the construction of railways and roads. Since this happened in Africa all through the twentieth century, the area from which export crops drained off increased all the time. (There was a corresponding expansion of the channel systems of individual exporters.) Normally, the product flow also thickened over the years, but it continued to converge in the port of shipment.

We now turn to situations where there is less agreement between the MCA and the geographical model. The single-port situation has, for instance, not been universal in Tropical Africa. The landlocked countries were the principal exception. In the colonial period almost all of them were linked by rail or river to one coastal colony and its port. After national independence, the exporters (and the governments) were free to choose between several coastal countries to reach an ocean port. In some countries, the external trade was indeed rerouted.[20] Another exception were the countries with two or more ports. In some cases, the hinterlands of these ports were quite separate; in others they overlapped.

The single-crop situation has also been exceptional. When there is more than one export crop, the question arises whether the product flows of all export crops follow identical channels. The geographical model suggests that this is so, but detailed research leads to a different answer. Thus, for palm oil, special measuring, testing, and storage facilities were required. Since not all exporters and intermediaries were prepared to invest in those, there were fewer channel systems for palm oil than for other crops and they covered a smaller area.

The single-exporter situation was common in the period after World War II, when in many countries one public exporter replaced a dozen or so private exporters. In this situation, the geographical model was most appropriate for describing the public exporter's channel system. In the other situation, that of several competing private exporters, each exporter had his own channel system—some large and well organized, and others small and weak. With only few exceptions, private exporters have had their headquarters in the ports because, as points of convergence, they bestowed a natural channel leadership on them. This was the best location for securing channel control.[21]

Glossary

cash on delivery (COD) at the PMP: The buyers' practice of promptly paying the growers when the latter deliver their parcels at the PMP.

contingency margin: Part of the exporter's *ex ante* trading margin. The exporter adds it to his normal trading margin so as to cover against an unexpected price fall.

direct trade: An upper section trading pattern in which the exporter buys directly from the growers.

exporter's interval: The interval during which the exporter is in possession of a particular quantity of the export crop.

gross trading margin: The margin between the trader's selling price and his buying price for a particular quantity of the product. In the case of an exporter it is desirable to distinguish between the *ex ante* margin (which the exporter wants or expects to earn) and the *ex post* margin (which he actually earns).

grower-exporter: An entrepreneur who is both a grower and an exporter.

indirect trade: An upper-section trading pattern in which the exporter buys from intermediaries, who in turn buy from the growers.

Inner Zone: The area between all the IMPs on the one hand and the port of shipment on the other.

intermediary's interval: The interval during which the intermediary is in possession of a particular quantity of the export crop.

intermediate marketing point (IMP): A point between the farms and the port of shipment where one or more intermediaries sell to the exporter(s).

internal marketing interval: The interval during which first the intermediary and then the exporter is in possession of a particular quantity of the export crop.

off-season [in export crop marketing]: The season during which there is no trade in the export crop. This is a period of idle resources and overcapacity in the enterprises of the intermediary and the exporter.

one-actor or grower-exporter pattern: An upper-section pattern with only one actor who is both a grower and an exporter.

Outer Zone: The area between all the PMPs on the one hand and all the IMPs on the other.

outgrower: A small-scale grower who has the right to sell his/her crop to a nearby plantation company which then processes the outgrower's output together with its own production.

parcel: Here defined as the quantity which the grower offers for sale at the PMP.

preharvest credit: Credit given by the intermediary (or the exporter) to the grower on condition that the latter sell his harvest to him.

primary marketing moment: The moment at which the grower sells his/her parcel.

primary marketing point (PMP): The point where the grower sells to the exporter or intermediary. The term also refers to the point where the farmer delivers his/her parcel to his/her cooperative.

processor-exporter: An entrepreneur who is both a processor and an exporter of a particular crop.

stationary buyer, mobile seller pattern: An upper-section trading pattern in which buying takes place at the buyer's store. The sellers (growers or intermediaries) have to travel and transport their produce.

three-actor pattern: An upper-section pattern with three actors: the grower, the intermediary, and the exporter.

trader-exporter: An exporter whose role is limited to trading.

two-actor pattern: An upper-section pattern with two actors: the grower and the exporter.

Chapter 7

Commodity TOMCs:
External Marketing—
The Exporter's Selling Conduct

Chapters 7 through 10 are devoted to the commodity TOMCs. The present chapter[1] deals with external marketing, which I have separated from internal marketing (in Chapters 8 and 9), but this separation entails two problems. First, internal marketing is not a common feature: it is absent in the grower-exporter pattern, but significant in the processor- and trader-exporter patterns. This is taken care of by Table 7.1, which shows how particular commodities tend to be associated with specific upper-section patterns. Second, since the separation of external marketing (downstream) from internal marketing (upstream) is somewhat artificial, some remarks about internal marketing have to be made in this chapter, even if they will have to be repeated later.

Commodities are supposed to require little or no active marketing management. As Terpstra (1983, p. 6) states, "One might say that commodities are sold and delivered, rather than marketed." This observation is true in the sense that promotion efforts are not required, but otherwise, it underrates the work of the commodity exporters, which includes challenging decisions. A detailed analysis of their *selling conduct* is useful for (1) the exporters themselves; (2) the supporting enterprises such as commercial banks and shipping companies; and (3) the governments of the exporting countries.

In the first part of this chapter, a new analytical framework is developed. In the second part, the historical evidence from Africa is reviewed. Before we turn to theory, we must identify the actors and the transactions.

TABLE 7.1. Commodities and Upper-Section Patterns

	grower-exporter	processor-exporter	trader-exporter
sugar	xx	outgrower	--
rubber	xx	outgrower	--
oil palm	xx	outgrower	--
arabica coffee	x	x	x
cotton	--	xx	x
robusta coffee	--	--	xx
cocoa	--	--	xx
oil seeds	--	--	xx

xx = well represented
x = rare
-- = absent

ACTORS AND TRANSACTIONS

In the field of external marketing, the exporters in Africa (who sell) and the importers in the countries of destination (who buy) are the main categories of actor. It is a special feature of the trade in commodities that there is no loss of analytical precision if importers are not further characterized. This is due to the exceptionally high degree of marketability of the commodities. Buyer search is no problem because there are plenty of buyers at any moment. Hence, the special motives of the individual buyer-importers (acquisition of raw material, stock holding, or speculation) and the corresponding conduct may be ignored.

As explained earlier, exporter and importer use standard CIF contracts, in which details of volume, quality, price, and approximate delivery date still have to be fixed. The preliminaries are as follows: the exporter keeps in touch with many importers; he may take the initiative and make offers or invite bids for particular quantities. If he is more passive, he may wait for bids from importers. Either course of action normally involves negotiation by telephone. When the parties agree, the terms of the contract are confirmed in two reciprocal messages embodied in telex prints.[2]

In their negotiations, the parties are always guided by the price quotations in the commodity exchanges. Most exporters depend on telephone contact with brokers, banks, etc., operating in the vicinity of the exchanges to obtain these quotations. Some large exporters have on-line access by means of the services provided by Reuter.[3] The price quotations are needed regardless of whether the exporter sells publicly in the exchange or privately outside it. In practice, nearly all the transactions discussed below are so-called private deals. The *effective price* of the contract must be analytically distinguished from the price quotations. There is usually a small margin (either a premium or a discount), which reflects the special quality of the country and the individual exporter's reputation.[4]

The availability of price quotations is assumed throughout this chapter. Hence, the analysis is not applicable to periods when the commodity exchanges were closed, such as during World War II.

A NEW ANALYTICAL FRAMEWORK

Two Views on the Commodity Exchanges

The commodity exchanges are primarily product markets, but they are also investors' markets. When seen as a product market, a commodity exchange is a market in which global supply and demand meet. Since competition is nearly perfect, international trade theory concludes that such markets favor the optimal allocation of resources: high-cost producers are replaced by low-cost ones—a process which occurs both in a national and an international context. As this may be considered well known, it is not further elaborated here.

When the commodity exchange is considered as an investors' market, the familiar categories of producer-seller and buyer-user are no longer appropriate. Instead, we encounter investors who sometimes operate as buyers and at other times as sellers. Consequently, temporary ownership is common. (Any reluctance toward such ownership is greatly reduced by the high marketability of the product.) The commodity exchanges (as well as the stock exchanges and the markets for foreign currency) have their own terminology for this situation. An actor who temporarily owns stocks has a long position. Such a position is liquidated by the sale of the tonnage involved. A short position, which exists when the actor's economic stocks are negative, is liquidated by the purchase of the tonnage concerned.[5] If the actor has a sound reputation, neither the actor nor his trading partners are worried about the short position.

Not only a long position, but also a short position exposes an actor to a *price risk*, for the value of his (economic) stocks fluctuates with the price quotations in the exchange. The holder of a long position fears a price fall and stock depreciation because it means that he loses money if suddenly forced to sell his stocks. (Of course, he desires a price rise, which permits making a profit by selling his stocks.) By contrast, the holder of a short position wants prices to fall. In textbooks, these situations are usually discussed with regard to speculators, actors who generally are not interested in holding stocks, but this chapter deals with actors who are unable to avoid holding physical stocks.

Since it is not easy to reconcile the two views on the commodity exchanges, writers usually take only one view. Thus, in theories of international trade, the view that the commodity exchanges are also investors' markets is nearly always omitted. This omission may go unnoticed when these theories focus on the long run because the complications caused by the investors' market are typically of a short-term nature. Governments also tend to omit the second view, largely because of the disagreeable connotations of the word "speculation."

Two Views on the Commodity Exporters

The two views just discussed determine our views on the exporters. In the first view (merely a product market), the exporter is seen as aiming at the highest effective price possible in his export contracts. It is this price that determines whether production in the country is viable. The relationship between (a) the effective export price and (b) the viability (or international competitiveness) is obvious when the exporters are also the growers. The situation is more complicated when exporters and growers are separate actors. We must then envisage the possibility that the exporters charge too high a trading (and processing) margin and thus discourage the growers. Much has been written about such "abuses" and the consequent distortion of incentives for the growers. However, this complication is irrelevant to the argument here.

In the second view (merely an investors' market), the exporter is primarily concerned about price risk. Since most exporters want to avoid this risk as much as possible, their *risk-avoiding conduct* in selling must be studied in detail. This will be done below. But this risk-averse attitude is not the only one. A few exporters display *risk-assuming conduct*, that is, they deliberately adopt either a short or a long position in hopes of a speculative profit. The fact that risk-avoiding and risk-assuming conduct are mutually exclusive complicates our analysis.

We may clarify the difference between the two views by using and extending the terms "price taker" and "price maker," which Scitovsky (1952, p. 6) introduced. When competition is perfect, both sellers and buyers are price takers, a term indicating that they have to take the ruling price as given. Scitovsky contrasted this with a situation of imperfect competition where either a seller or a buyer has some control over the price, thus enabling him/her to "make" the price—he/she is a price maker. In international trade theory, it is generally argued that the exporter of commodities and, as a corollary, the exporting country are price takers. This conclusion is reinforced by the SCP (Structure, Conduct, Performance) analysis, which postulates that the conduct of the actors in a market is determined by the structure (degree of competition) of the market (Scherer and Ross, 1990). The conduct of the commodity exporter is therefore supposed to be severely restricted, if not completely passive. As a result, many economists have not bothered to study the exporter's selling conduct, assuming that it offered no intellectual challenge and had no effect—a most unfortunate omission.

The new hypothesis that I advance is that the exporter is a *moment maker* and that moment-making policies are profitable. Within an *effective time range*, the exporter is, under certain conditions, free to choose the moment of sale. True, as soon as the moment is fixed, the exporter is a price taker, but as moment making increases the range of opportunities, it may, on *a priori* grounds, be assumed to be beneficial for the exporter. To see how this works in practice, we must explore the exporters' freedom to choose their moment. This exploration involves at least three factors: (a) the agricultural cycle and the effective time range; (b) the default risk and the exporters' reputation for never defaulting on their contracts; and (c) the exporters' attitude toward risk. We shall see that some exporters confidently resort to moment making in order to avoid risk, and that others, equally confident, use moment making to engage in risky policies.

The Agricultural Cycle and Physical Stocks

We first examine how the agricultural cycle influences physical stocks in the upper section in Africa. We do so by means of a diagram for a seasonal crop.[6] The stock cycles in Figure 7.1 are drawn so as to emphasize similarities between the three parts of the diagram. Thus, the cultivation intervals between points A and B are identical. During the cultivation interval, there are *stocks in sight* which are also exposed to the price risk. The price quotations during this interval are relevant because, even at this early stage, they form the best indication of profit prospects. The straight line from point A to point P is a simplification, because in reality, neither the volume nor the value of the growing crop is likely to exhibit such a neat line.

In all three parts, the crop is harvested at point B and shipped at point F. (To simplify the argument, we assume that these activities can be completed in one day.) The length of the *postharvest interval* between points B and F (during which domestic transport, storage, and possibly processing and marketing take place) is not uniform. The shortest interval is found in Figure 7.1a, where coordination is superior. In Figure 7.1b, processing is responsible for a long interval. As the horizontal line between points P and Q shows, the physical stocks are constant during the postharvest interval.[7] Throughout this interval, physical stocks are exposed to the price risk. Since we assume for the time being that the stocks are sold at point F, exposure to the price risk ends at that moment or, more accurately, the price risk is shifted to the buyer-importer.[8] Point F is the moment of truth for the exporter: shaky profit prospects turn into firm realized profits (or losses). During the off-season between points F and A, there are no physical stocks and there is no exposure to the price risk.

FIGURE 7.1. Annual Commodity Stock Cycles

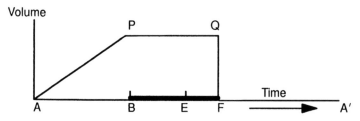

7.1a. Annual stock cycle of a grower-exporter

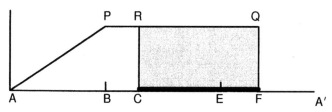

7.1b. Annual stock cycle of a processor-exporter and his suppliers

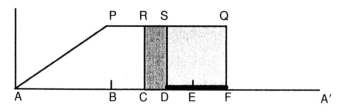

7.1c. Annual stock cycle of a trader-exporter and his suppliers

Point A = sowing
Point B = harvest
Point C = primary purchase
Point D = intermediate purchase
Point E = arrival in port warehouse
Point F = shipment
Point A' = sowing (next season)

━━━━ = central interval

Actor versus Sector Perspective

The diagram is drawn in such a way as to permit both a sector and an actor perspective. In Figure 7.1a, the stock cycles of actor and sector coincide because there is only one actor, but in Figure 7.1b, there are two actors: the grower who owns the stocks to the left of the line CR and the processor-exporter who owns those to the right. In Figure 7.1c, there are two vertical lines: line CR separates the stocks of the growers from those of the intermediaries and line DS does the same for intermediaries and the trader-exporter. Corresponding statements can be made about the long position of each actor, provided that economic stocks coincide with the physical ones as they normally do. This terminology can be extended to the sector: it is possible to say that, in all three cases, the sector (and the country) has a long position during the whole interval between points A and F.

From the sector perspective, any price fall (and the concomitant stock depreciation) during the entire interval between points A and F is relevant. However, the actor has a narrower view. He is hurt only by price falls which occur during the periods when he holds stock. Thus, for the actors in Figures 7.1b and 7.1c, it matters a great deal whether prices begin to fall during the cultivation interval or a later interval. If in Figure 7.1b, for example, the fall in prices occurs (and comes to an end) before point C, a processor-exporter can avoid losses by setting the buying price low to be in line with the world price at point C. (This may be an unpleasant surprise for the growers, who do not normally discover adverse price movements until they sell.) Conversely, if the world price begins to fall after point C, the growers are not hurt at all. Clearly, the periods of exposure to the price risk do not coincide for the actors.

In Figures 7.1b and 7.1c, the financial results of the exporter must be distinguished from those of the other upper-section actors and the sector as a whole. Thus, if the accounts of a processor-exporter show a profit, we should not conclude that the growers' operations have also been profitable. In a price fall scenario like the one just sketched, it is possible that the growers suffer a serious loss and the exporter makes a profit. It is even possible that the sector (taken as a unit) has a negative result for the season. The situation is even more complex in the situation of Figure 7.1c. Trader-exporters may, for instance, suffer losses in the same year that intermediaries earn handsome profits.

It is also essential to recognize that the actors are not all equally vulnerable to the price risk. The potential depreciation loss should be related to the actors' annual profit margin. If the margin is narrow, the actors' vulnerability is greater. This is particularly true if the gross margin is narrow (as is the case with the trader-exporters and the intermediaries) because not only the profits, but also the liquidity position, is threatened by a price fall.

Three Selling Intervals

We assumed above that selling and shipping always coincided, that is, the moment of sale was determined by the moment of shipping, which in turn was determined by climatic and technical factors. We must now abandon our initial assumption and investigate the freedom of moment making. Taking the diagram as our starting point, we divide the exporter's effective time range into three intervals: (a) an early interval (to the left); (b) a central interval; and (c) a late interval (to the right of point F). Sales during the early interval always lead to a short position because the exporter does not yet own the crop. The central and late intervals correspond to a long position (see Table 7.2).

We first consider sales during the central interval. The exporter has to take two risks into account: the price risk and a default risk, that is, the risk of not being able to deliver in accordance with the terms of the sales contract. The default risk is not constant but, as a rule, decreases, as the product moves from left to right. Allow me to illustrate this for internal transport. After the product has arrived in the port warehouses (at point E in all three cases), the risk of transport delay is reduced to zero. A similar reasoning applies to processing, such as cotton ginning. At the beginning of the central interval, a cotton ginner owns (unmarketable) raw cotton. Within narrow limits, a cotton ginner is able to estimate the tonnage of cotton fiber he will obtain by ginning. For this tonnage, he considers himself exposed to the price risk. If the cotton ginner enters into a forward sales contract to end this risk, he exposes himself to the risk of default, particularly if his installations are old or poorly

TABLE 7.2. Aspects of Moment Making by Commodity Exporters

Name of Interval (1)	Exporter's Position (2)	Default Risk (3)	Attitude Toward Risk (4)	Moment Making (5)
early	short	serious	assumption	beneficial in the long run
central	long (hard to avoid)	moderate	avoidance	beneficial in the short run
late	long (easy to avoid)	negligible	assumption	disappointing in the long run

maintained and liable to break down. These illustrations indicate that much may go wrong.

Cautious exporters tend to sell late, i.e., during the interval between points E and F. As a rule exporters are drawn in two directions. If they wish to avoid the default risk, they sell after point E, but if they are worried about the price risk, they want to sell before point E.

One more point should be made here. If the moments of sale and shipment are not far apart, the exporters are guided by the spot prices, but if there is a long interval, the futures prices are more relevant to the exporters. Here, a new risk enters. It is inherent in futures market signals that they may turn out to be wrong later. It is only at point F that an exporter may acquire the wisdom of hindsight—and may feel regret. If the spot price at that moment turns out to be much higher than the signals predicted, the alternative policy (postponing sales until the time of shipment) would have been better.

We now turn to selling during the late interval. It is nearly always possible for an exporter to avoid such sales because neither agricultural nor logistic factors compel him to sell so late.[9] The default risk is negligible for all three categories of exporter. As to an exporter's conduct, we simply note that he remains exposed to the price risk. Either the exporter is not worried about this risk or expects a price rise and hopes to profit from it.

Finally, selling during the early interval must be considered. This normally involves a long period between the transaction and the shipment. Here, we briefly note that even more may go wrong than in the central interval. Indeed, the longer this period, the greater the chance of unexpected problems. Early sales are also associated with a short position and reveal risk-assuming conduct. A more detailed discussion, taking each category of exporter separately, follows in the next section.

Is a buyer-importer on the other side of the ocean prepared to accommodate the exporter's wishes with regard to the moment of selling? In particular is the buyer-importer willing to agree to early sales? In trying to answer these questions, we must consider an importer's time perspective. He is familiar with conditions in the countries of shipment and knows the points B and F for each country he buys from.[10] He knows whether the approximate date of shipment in the contract is realistic. The importer can derive from it the date of delivery in the port of destination, which, of course, is his primary concern. Since there are many exporting countries, the importer takes the moment of transaction as an independent point of reference. Starting from this moment, an importer measures the interval until shipment. Depending on its length, he distinguishes between prompt, nearby, and distant shipment. (Prompt roughly corresponds to transactions during the interval between points E and F or to the right of F; nearby refers to the interval between points B and E; and distant refers to the interval between points A and B.)[11]

An importer who is about to enter into a contract with distant shipment[12] primarily considers the default risk. He does not enter into a contract unless he considers the default risk acceptable. Here, the reputation of the exporter plays an important part because it relieves the importer from the need to make an independent estimate of the factors which might lead to default on a particular contract.

As soon as the importer has entered into a purchase contract, the price risk is shifted to him, even if the physical stocks are still in Africa. Of course, the importer fears the price risk and is reluctant to bear it. However, an importer's reluctance is reduced when there is a terminal market for the commodity involved because he can then offset this risk by hedging. For such an importer, it is important to know how far ahead the futures prices are quoted. For tree crops, futures prices tend to cover the longest period: up to two years. To sum up, not all importers are willing to accept distant shipment, but there are enough of them to accommodate exporters who want to use their moment-making freedom.

The Default Risk of a Short Position

While the default risk of a short position is always serious, it is not of the same kind for each of the three categories of exporter. For grower-exporters, the default risk is primarily the risk of partial failure of the crop growing on their lands. If they wish to sell early, it is important for them to inspect the standing crop regularly and to have up-to-date harvest estimates. Even so, prudent exporters do not sell more than perhaps 70 percent of the estimated volume during the interval between points A and B. (If the harvest turns out to be equal to the estimate, the remaining 30 percent can be sold later.) In the case of tree crops, the grower-exporters may sell before point A, even several months before point A.

The default risk of a processor-exporter, for example, a cotton ginner, consists of two elements: the cotton crop of the farmers from whom he intends to buy may partly fail, and these farmers may refuse to sell their cotton to him. For a ginner who wants to adopt a short position, it is therefore desirable (a) to monitor the cotton crop during the growing season and (b) to be the sole buyer and only outlet for the growers. The ginner may then count on captive supplies (see Chapter 8) and is able to enter into early sales contracts.

When a trader-exporter adopts a short position, his default risk consists of three elements: the crop may fail, the growers may refuse to sell their crop to the intermediaries, and the intermediaries may refuse to resell the crop to him. As a rule, the more successive actors there are in the channel, the greater the default risk of the exporter. However, this risk is reduced when the exporter has *de facto* control over the upstream actors and successfully manages his relationships with those actors and establishes certain claims on supplies from them. This may be a matter of captive intermediaries and/or captive growers (see Chapter 9).

The Exporter's Attitude Toward Risk

The attitudes of exporters vary from risk avoidance to risk assumption and this is related to the timing of their sales (see Table 7.2). We may be brief about the risk-averse exporters who sell during the central interval. They minimize the default risk, notably by means of managerial alertness and competence. And they shift the price risk to the importer by means of forward sales.

It is further possible to be brief about the risk-assuming exporters who sell during the late interval. Their decision to do so is based on their expectation that prices will rise more than the market expects. They do not *follow the market*, but trust their own intuition, rather than the price quotations. These exporters may properly be called bull speculators. This term puts them on one line with the bull speculators in the commodity exchanges. This is appropriate, because, like those in commodity exchanges, they are guided by price expectations only. Indeed, since the default risk is negligible, it plays no role in the exporters' considerations. I have serious doubts about the wisdom of such late sales because exporters in the periphery of the world are informationally handicapped.

I have a positive opinion about the risk-assuming exporters who sell during the early interval. I interpret their conduct as follows: they consider themselves the prospective owners of part of the new harvest. A particular exporter may thus see himself as the future owner of X tons of cocoa. Being quite sure of becoming the legal owner later in the season, he has no hesitation to sell these X tons now. Whether the exporter actually does so depends on the futures prices corresponding to point F. If these are favorable, he will enter into a private forward sale for X tons. The exporter engages in *tactical speculation* because he feels that a higher price can be obtained by selling now (and shipping later) than by postponing the sale to a date close to the date of shipment.[13]

Three supplementary points must be made. First, what I have called favorable futures prices corresponds to contango situations in the commodity exchanges (see Chapter 5). During (irregularly occurring) contango situations, the futures prices are higher than the spot prices. The resulting contango premium of distant sales over prompt sales acts as an incentive for the exporters to adopt a short position. Since in practice, this premium has an upper limit (which depends on carrying costs in the ports of destination), the exporters' profits from tactical speculation are limited, but it must be pointed out that profits are systematically greater for the more distant shipments.[14] It does not seem correct to me to call these exporters bear speculators, although this term has sometimes been used by exporting companies (Fieldhouse, 1994, p. 114). In the same way, I reject the idea that early sales are the mirror image of late sales. There is a fundamental difference. In early sales, the exporter has to consider both the price and the default risk, while in late sales, only the price risk operates.

Second, whether it is wise for the exporter to engage in tactical speculation depends on the default risk of a short position (discussed above). Here, I add that the exporter has an emergency solution at his disposal. If the supplies the exporter was counting on are not forthcoming, he may buy supplies elsewhere and use these to fulfill the contract. This is not the solution that the parties to the contract had in mind, but it prevents default. In this exceptional case, the exporter is not a seller in the commodity market, but a buyer! In this way, the exporter benefits from the fact that the commodity exchange is an investors' market.

Third, the exporter makes his profit calculations on the assumption that he will be able to acquire ownership on normal conditions. The exporters' anticipated profits may be eroded if these conditions unexpectedly change. Thus, a grower-exporter may be disappointed when wages for agricultural labor increase. Similarly, trader-exporters may lose money if competition forces up their buying prices. Obviously, exporters have a strong incentive to restrict competition among themselves. When exporters engaged in schemes to restrict competition in the past, the motive usually attributed to them was that of forcing down the buying price. The desire to do so was supposed to be continuous. In the analysis presented here, the motive is different (captive supplies) and the desire varies in intensity over time, being strongest in contango situations.

Macroeconomic Aspects of Moment Making

Our microanalysis of the exporters' moment-making policies has led to the following seemingly inconsistent conclusion: risk avoidance is beneficial in the short run, but risk assumption (on early sales) is beneficial in the long run (see Figure 7.2, Column 5). Does a macroeconomic perspective throw further light on this remarkable conclusion?

The macroeconomic merit of the risk-avoiding exporters is that they minimize losses due to stock depreciation, not only for themselves, but also for the country from which they are exporting. Even if their conduct is, in principle, approved by an African government, two problems must be recognized. First, selling during the central intervals leads to the bunching of export sales, which may amount to disruptive off-loading on the world market and may thus depress prices.[15] Second, an exporter only minimizes the price risk for himself, not for any upstream actors from whom he buys. In the eyes of the growers in Figures 7.1b and 7.1c, the exporter acts egoistically because he passes the short-run movements of the world price on to them.

African governments also face more basic questions: Is it always desirable to play safe and to shift the price risk to importers abroad? What do importers charge for taking over this risk? Is the average price obtained by the countries of origin systematically lower because of the transfer of the price risk? In recent years, a great deal has been written about *price risk management*, that is, about policies and instruments to shift the price risk to another actor. Instead of the forward sales contract which had been the principal instrument for many decades, new instruments such as options were developed by financial institutions in the 1980s. As the fees charged for such options have made the cost of price risk transfer transparent, there is now a better chance of estimating the national costs than in the time of forward sales. Clearly, these costs could be saved if the exporting country was able to assume the price risk. However, because of the unpredictability of commodity prices, the savings would only be realized in the long run. What comfort does the long run offer if all the blows come at the beginning?

It seems that governments have rarely distinguished between risk avoidance and risk assumption and have seldom expressed a preference. Instead, governments were carried along by the current mood of optimism or pessimism among the exporters. When we examine the mood of the last ten years, pessimism has clearly predominated.

The whole body of writing on price risk management is, for instance, based on the view that Africa's financial shoulders are too weak and that the price risk should therefore be shifted to strong shoulders elsewhere. While pessimism is presently appropriate, times may change and the macroeconomic merits of risk assumption may come in view again.

Anticipating a time when governments again accept risk assumption as a proper objective, I want to specify three consequences. First, the government must support strong export organizations. Only strong exporters (whether public or private) are able and willing to take a long-run view. A strong channel leader is needed to have informal control over the upstream actors (which is needed to reduce the default risk of a short position). Second, the government must invest in internal transportation in order to facilitate a smooth flow of the product and thus reduce the default risk. Third, the government should not rely blindly on the discipline of the market in ensuring the viability of the export sector. Risk assumption may lead to depreciation losses and windfall profits, either of which distort the relationship between total profits and operating costs. In fact, unlucky moment making may liquidate an efficient private exporter, while lucky timing helps a high-cost rival to survive.

THE HISTORICAL EVIDENCE

In reviewing the historical evidence about the exporter's selling conduct we may distinguish three periods: 1880 to 1940 when private exporters were the rule; 1940 to 1986 when public exporters predominated; and 1986 to the present when many public exporters were replaced by private ones. It must be pointed out from the beginning that the evidence, in spite of the long period, is extremely scarce. Moreover, there is less information available for the grower-exporters than for the trader-exporters and the processor-exporters.

The Period Before World War II

Substantial evidence on the pre-1940 situation is only available for the trader-exporters in British West Africa. An important source is the Nowell Report (1938) on the cocoa exporting companies in the Gold Coast (now Ghana) and Nigeria. In the 1930s the majority of these companies tried to keep their economic stocks of cocoa close to zero. This policy corresponds to selling at point D in Figure 7.1c and was known in the 1930s as the policy of "squaring one's books." The main problem besetting this policy was to ensure rapid and reliable communication between the company's up-country buyers and the general manager in the port of shipment. The general manager had to make sure that the up-country purchases of any particular day were matched by international sales during the same day.[16] This policy indicates that the exporters were greatly worried about losses due to stock depreciation.[17]

These cocoa exporters operated from an actor perspective. In the 1930s they only thought of their own interests and were unwilling to make sacrifices for the survival of the sector. (If the sector faltered, it was seen as the task of the government to tide it over during the bad years.) The Cocoa Buying Agreement of 1937 (further discussed in Chapter 9) may be seen as a formalization of the policy of selling at point D.[18]

One cocoa exporter, the United Africa Company (UAC), adopted a different policy. It was compelled to do so partly because it was a large exporter in West Africa (accounting for some 50 percent of West Africa's cocoa exports) and partly because West Africa's share of world cocoa supplies was very high. Since UAC was a "megaexporter," it would depress prices if, like its rivals, it sold at point D.[19] Instead, UAC had to spread its selling contracts over many months. Each year, it began with a short position which gradually changed into a long one, with unsold stocks peaking in March.[20] Because of its solid reputation, UAC could easily engage in distant sales.[21]

The Public Exporters 1940 to 1986

During World War II, the British authorities established public trading organizations to cope with the emergency war situation. These were the forerunners of the British-type *export marketing boards (EMBs),* set up in the late 1940s. In the 1950s the French authorities established *Caisses de Stabilisation* (price stabilization funds) for cotton, cocoa, coffee, and oil seeds (Vloeberghs, 1956). A crucial element of both the EMBs and the *Caisses* was that the price risk was transferred from the private to the public sector. Within a few years, academic and political interest in the price risk faded away.

A conspicuous difference between the EMBs and the *Caisses* was that the former operated as state enterprises, while the latter were merely funds, leaving the actual export function to licensed private exporters. However, for the purposes of this chapter, this difference is of minor significance. Here we focus on exporter's conduct and how the design of the two types of organization affected it (see Table 7.3).

In Figure 7.1, exposure to the price risk is shown to depend on the agricultural cycle. As a result, only the short-run price fluctuations seem significant, but this view is too narrow. In fact, those studying the commodity price movements have always tried to distinguish long-run and short-run components. Since this seemed impossible in the interwar period, governments were then generally unwilling to bear the price risk. At most, they were prepared to promise financial support to private exporters in case world prices fell below a certain floor.

In the 1940s, many economists came to a new interpretation of the price fluctuations. They held that the long-run component was largely determined by the business cycle. It was therefore considered possible to achieve some degree of stability by means of *countercyclical policies*. Westcott (1987) shows that the British government at first pursued international price stabilization schemes. When support for intervention in the commodity exchanges (by means of buffer stocks) dwindled, a country-by-country solution was adopted to ensure at least domestic stability. To the laymen the countercyclical policy was presented as one of evening out revenues over a period of some ten years by means of stabilization reserves, which were fed in good years and tapped in bad ones.[22]

Table 7.3 shows how far-reaching the creation of the EMBs was. Five important objectives were simultaneously realized. The French approach was different: the objectives remained separate and the institutions were more flexible and started at different moments. Thus, the French had already accepted the principle of transferring

TABLE 7.3. Price-Risk-Bearing Public Bodies in Tropical Africa Since 1930 by Objectives, Design, and Year of Introduction

Objectives	British Design	French Design
1. Long-run price risk, partly and/or conditionally, borne by public sector	EMB (1947)	*Caisse de Soutien* (1930)
2. Long-term financial reserves (separate from government budget)	EMB (1947)	*Compte hors budget* (1947)
3. Short-run price risk borne by public sector	EMB (1947)	*Caisse de Stabilisation* (1956)
4. Direct protection of growers against price risk	EMB (1947)	---
5. Legal basis for monopsony	EMB (1947)	---

Notes: * *Caisses de Soutien* (price support funds) for cotton were established in 1930 in Chad and Ubangi-Shari.
* The first EMBs were the cocoa boards in the Gold Coast and Nigeria.
* A *Compte hors budget* (long-term account kept separate from the government's budget) for cocoa was established in Cameroon in 1947.
* The existing *Caisses de Soutien* and the *Comptes hors budget* were incorporated into the new *Caisses de Stabilisation* in 1956.
* In several francophone countries, an *Office de Commercialisation* was established in the 1960s. It was grafted onto the French design and replaced the existing *Caisse de Stabilisation*. Its responsibilities were increased to cover objectives 4 and 5.

the price risk to the public sector around 1930. However, it was a limited transfer. A public fund, a *Caisse de Soutien* (price support fund), was created to make support payments to licensed exporters, but only when world prices were very low.[23] The *Caisses de Stabilisation,* established more than 20 years later, marked both an acceptance of the countercyclical approach and the systematic transfer of the short-run price risk to the public sector.[24]

The last two objectives of Table 7.3 refer to the domestic or upstream aspects of an EMB. It is significant that EMBs were only set up in sectors where growers and exporters were separate actors, that is, in the sectors illustrated by Figures 7.1b and 7.1c.[25] In the British design, it was not enough to protect only the exporters against the price risk: the growers had to be protected too. Only a public institution was sufficiently impartial to pursue the interests of the sector as a whole. Actual protection was effected by the so-called *producer price*, the up-country buying price fixed by the EMB and announced early in the season. For annual crops such as cotton and oil seeds, the ideal moment to make the announcement was prior to sowing time (point A in Figure 7.1). For tree crops, it was normally done later, usually just before harvesting began. We note for later reference that the EMB's exposure to the price risk began on the day on which the producer price was announced.

The EMB designers held the view that protection of the growers was impossible without the creation of a monopolistic public exporter—a view not followed by the French. As early as the 1950s, the buying monopoly or *monopsony* was attacked by British economists, for instance by Bauer (1954).[26] Unfortunately, Bauer considered only the upstream implications of having strong EMBs. He expected (correctly, as it turned out) that governments would abuse the EMB's strength to lower the producer prices at the expense of the growers. However, Bauer failed to see the downstream advantages of a strong EMB!

In the French design, direct protection of the growers was not envisaged. The French relied on competition among the licensed exporters to ensure that the benefits of the (annually fixed) port-of-shipment prices were felt up-country. The new African governments, less optimistic about competition, decided to modify the *Caisse* system by introducing up-country producer prices. In many countries, this was done in the 1960s. It was done in Côte d'Ivoire and Cameroon in 1965 (Ramboz, 1965; Assoumou, 1977). This step brought the *Caisse* more in line with the EMBs. In some countries a further step was taken: the licenses of the private exporters were withdrawn and the *Caisse* became a public exporter. These steps were usually accompanied by a change of name: the *Caisse* became an *Office de Commercialisation*.

The Selling Conduct of the Public Exporters

Little is known about the selling conduct of the early EMBs. As long as the system of bulk buying by the British Government operated (roughly until 1954), the EMBs were severely restricted in their freedom to sell.[27] Afterward, the sales function seemed unimportant. I assume that this was due to the then-dominant view that commodity prices were determined by the business cycle and could effectively be stabilized. In the 1960s, this view was imperceptibly replaced by the view that prices were basically unpredictable. Academics then began to recommend *export-parity pricing*, that is, the EMB should annually fix its producer prices in line with current world prices. If this policy was followed, the EMB would normally break even each year. True, the long-run price risk was no longer covered, but short-run protection for the growers continued.

It is plausible that export-parity thinking revived attention for short-run price movements. What was more logical for an EMB (and an *Office de Commercialisation*) than to sell its next harvest forward at about the moment that it fixed the producer price? The EMB's selling price could thus be linked to its buying price, not only value-wise, but also time-wise.[28]

Four remarks about this hypothesis are in order. First, for the annual crops it was unwise to sell much at this early stage because the size of the new harvest was still uncertain.[29] Second, there was no serious default risk in early sales because the EMB's statutory monopoly, turned the standing crop into captive supplies.[30] Third, because of its monopoly, a cocoa or coffee EMB with a sound international reputation was able to sell its product more than a year before the anticipated date of shipment. In fact, such very early sales, a special form of tactical speculation, were often highly profitable. Fourth, since the export quotas of the International Coffee Organization (ICO) were based on quarters (rather than on full years), the coffee EMBs were compelled to spread their shipments over the year. Presumably, this led to a certain spreading of sale contracts, first in the coffee trade and then in other trades.

In the francophone countries, there was a similar revival of interest in the opportunities for tactical speculation. The *Caisse de Stabilisation* in Côte d'Ivoire began to export on its own account around

1970. It made so-called direct sales (Delaporte, 1976).[31] What was not sold by the *Caisse* continued to be sold by the licensed exporters. However, the *Caisse* selected the plums of the pudding. It was widely rumored that president Houphouet-Boigny was active in these direct sales. Whether this is true or not, the top managers of most public exporters were reluctant to engage in tactical speculation unless they had political backing.

Until about 1970, the EMBs were not greatly worried about the *currency risk*. Normally the contract price was expressed in a major currency such as the U.S. dollar, which was quite stable until 1971. A prudent EMB could seek coverage against the currency risk in one of the terminal markets for foreign exchange. The coverage must have corresponded to the period of risk exposure, which ended when the importer took delivery of the consignment and paid. A different solution was provided by the price-to-be-fixed contract, which granted the EMB the right to fix the price at the moment of its choice on the basis of an agreed-upon formula.[32] One advantage of such contracts for the EMB was that the currency and price risks remained combined until the decisive moment. Yet another solution was the option (already mentioned). The option giver guaranteed the EMB a minimum price (which was a protection in case the world price fell), but allowed it to benefit from a price rise.

We end this section with some information on the cocoa EMBs. (Data on them have been more abundant than on the other EMBs—a parallel with the 1930s.) Cocoa was the first commodity for which free trading resumed after World War II. This happened in 1947 when the New York cocoa exchange reopened. The response of the cocoa EMBs in British West Africa (i.e., in the Gold Coast and Nigeria) was most interesting. Their marketing companies in London pursued a joint selling policy based on their status as a megaexporter. Before the war, UAC had handled approximately 50 percent of West Africa's cocoa; now, nearly 100 percent was being handled by these two EMBs. Their main London-based seller, Eric Tansley (later Sir Eric), had been a top UAC cocoa employee before the war. It would not be far-fetched to suppose that he applied the lessons of the 1930s within the new public export organization.[33] While he was in charge (until 1961), he used his position to stabilize the international cocoa market.[34]

A remarkable event occurred during the first decade after most African countries had gained independence. The Cocoa Producers Alliance (CPA) attempted in 1964 to drive up world prices by slowing down or stopping shipments of cocoa from the six member states: Ghana, Nigeria, Côte d'Ivoire, Cameroon, Togo, and Brazil (Assoumou, 1977).[35] Since the cocoa EMB in Ghana was a major actor, the whole scheme may well have been President Nkrumah's brainchild (Pedler, 1979, p. 134). Also Hanisch (1980) interprets it as the international version of the cocoa hold-up in the Gold Coast in 1937/38, which had been a political and economic success for Africans. However, it did not work that way in 1964/65. Prices did not go up; in the end, the CPA countries were obliged to export at very low prices. In theoretical terms, the cocoa EMBs in the CPA countries had adopted an abnormally long position. The failure of this strategy seems to have deterred the EMBs for many years afterward from postponing sales to a moment to the right of point F.

Private Exporters Since 1986

In the 1980s, many African governments adopted structural adjustment policies, which included the privatization of the EMBs[36] and their replacement by new private exporters. One of the highly publicized and sweeping policy changes in Nigeria was the abolition of all its marketing boards in July 1986,[37] which is a reason why I have taken the year 1986 to demarcate this third period.

In Table 7.4, the new private exporters are compared with the earlier EMBs. In principle we cannot go farther than comparing the institutional arrangements and we must be wary of interpretations which attribute recent changes to privatization without considering other possible explanatory factors. When we adopt a market theory approach and confine our attention to internal (or upstream) marketing (points 1 and 2), the generally accepted conclusion is that the farmers are better off under private exporters. This is based on a firm belief in the superiority of private enterprise, whose operating costs are expected to be lower than those of the public EMBs. EMBs, it was assumed, had not been overly concerned about efficiency, sheltered as they were by their monopoly.[38] In essence this was the argument advanced by Bauer (1954) 30 years earlier. It became a basic line in structural adjustment.

TABLE 7.4. Comparing the Institutional Arrangements in a Commodity TOMC Before and After Privatization

	One Public EMB	Many Private Exporters
1. internal marketing characterized by	monopoly	competition
2. internal marketing costs	higher	lower
3. channel systems	one large	many small
4. control over upstream actors	strong	weak
5. exporter's international reputation	good	poor
6. timing of sales	spread out	concentrated
7. tactical speculation	possible	impossible
8. FOB price	higher	lower
9. effect on farmers' income	uncertain	uncertain
10. sector perspective	decisive	abandoned
11. farmers exposed to price risk	no	yes

When we adopt the MCA approach and focus on external (or downstream) marketing (points 3 through 8), we must turn to the channel systems of the private exporters. If exporters operate on a small scale and have no up-country branches, they cannot influence the buying operations up-country, e.g., with regard to quality. (This is confirmed by the drop in quality of cocoa exported from Nigeria after 1986.) Such small exporters are unable to make sales outside of the interval between points E and F: they sell from port stocks with prompt shipment contracts and last-moment sales. This situation is described in an UNCTAD report (UNCTAD, 1994), which concludes that contract prices have declined. Indeed, FOB prices have probably been lower after privatization.

Thus, from the MCA perspective, the privatization of the EMBs has definite drawbacks. Certain assets which existed under the EMB and were beneficial for the whole sector have lost part of their value, and certain opportunities have disappeared. First, insofar as an EMB had a solid reputation among importers abroad, this good-will disappeared with the dissolution of the EMB. Since the reputation of the new private exporters does not compensate for this, there is a national loss of goodwill. Second, as the moment-making freedom depends on reputation, this freedom and the opportunity for tactical speculation practically disappeared. Third, the skills acquired within the EMB were diffused and dislocated. Only some of these skills are presently employed in the sector. The overall negotiating position of the sector vis-à-vis the importers abroad has deteriorated.

The crucial question is now whether the positive effects of privatization (as postulated by market theory) outweigh the negative effects (as postulated by MCA). I myself am not convinced that the privatization of the EMBs was beneficial for the farmers (see point 9 in Table 7.4). However, a further point should be made. Even if it could be established that privatization raised farmers' incomes, the uncertainty about prices during the cultivation interval is a disadvantage, albeit one that is difficult to express in monetary terms.

Problems Caused by Privatization

There are now serious misgivings about the wisdom of the privatization drive of the 1980s, as is evident from various proposals to remedy recent problems. Some proposals aim at strengthening the private exporters and upgrading their performance. First, commercial banks have begun to extend guarantees to exporters to enhance their international reputation. Second, in several countries, the exporters have established an exporters' association, which intends to raise the reputation of its members by having stricter codes of conduct and by expelling dishonest members. (However, reputation is not only a matter of honesty; it also requires experience and managerial competence.) Third, many training courses on international commerce, including price risk management, have been provided.

Many governments are worried that the new exporters are inexperienced in managing the price risk and may make mistakes that are harmful for the sector as a whole. These problems are addressed in a recent World Bank publication (Claessens and Duncan, 1993). This book examines the instruments that are internationally available. Options are now often preferred to hedging operations because access to them is generally easier for exporters.[39] Attention is further given to the extent to which large private businessmen protect themselves, the farmers, and the economy as a whole against price risks. This question has been studied for the coffee sector in Costa Rica (Claessens and Varangis, 1993).

The discussions on price risk management may well lead to a reassessment of the EMBs. During the privatization drive, they were portrayed as passive actors with regard to risk management. Without the spur of competition, they supposedly did not worry about efficiency, and similarly, without the threat of bankruptcy, they did not worry about price risks. But is this picture correct? Further research is highly desirable.

Another category of remedial proposals concerns the farmers. Farmers became exposed to the price risk when the EMBs were abolished. Before privatization, the farmers had been protected during the cultivation interval. Now, they remained in uncertainty about the outcome of their efforts until they sold to the exporters or intermediaries.

The question is whether and how farmers can be protected against the price risk. Some people have looked at the example of farmers in the United States, notably grain farmers, who have worked with preharvest hedging to cover the price risk during the cultivation interval (Kohls and Uhl, 1990, pp. 343–334). However, most of the farmers in Africa are too small and too unsophisticated to take the initiative in risk management. That is why the UNCTAD report just mentioned examines the possibility of national risk intermediaries. Private traders (which correspond to our trader-exporters and processor-exporters) can indeed act as risk intermediaries, but there is no indication that they will do so during the cultivation interval. Here, the lessons from the 1930s are relevant: an exporter does not engage in risk management until he buys and the farmers are left high and dry.[40] Another actor that has been proposed as a risk intermediary is the farmers' cooperative. Once again we are reminded of the discussions of the 1930s, when some observers envisaged a national cooperative as the ideal exporting organization for the cocoa sector in the Gold Coast (Nowell Report, 1938). One merit of this chapter is that it helps to compare the problems of the 1930s with those of today.

Current discussions on external marketing seem one-sided to me because of their preoccupation with the price risk. The analysis in this chapter suggests that the default risk is at least equally important. I also contend that default on an export contract should be perceived in a specific way. In Africa, protection against such default is primarily sought in managerial competence rather than in financial standing. Because of this view, I expect the reemergence of large exporters in Africa. In some countries, these will be foreign companies which now operate in the lower sections of the TOMC. Elsewhere, there will be large local exporters, possibly supported by politicians.

Glossary

countercyclical policies [with regard to commodity prices]: Policies based on the assumption that the movements of commodity prices are primarily determined by the business cycle.

currency risk: The risk that the currency in which the contract price is expressed will depreciate in the near future. Exporters are exposed to the currency risk from the moment they enter into the export contract until their importer pays.

effective price [of an export contract]: The actual price agreed upon in the contract.

effective time range: The period during which an exporter is able to choose the moment of sale. The range consists of three intervals (early, central, and late) as shown in Figure 7.1.

export marketing board (EMB): A state trading enterprise, established to buy export crops from (mainly small) farmers and to market them abroad. An EMB usually possessed a buying and a selling monopoly. The protection of African farmers against (a) price fluctuations in the world markets and (b) exploitation by exporters and intermediaries, were major objectives of the British-designed EMBs in Tropical Africa.

export-parity pricing: The exporter's practice of setting (upstream) buying prices, notably producer prices, in the light of (downstream) export prices in such a way that the exporter's *ex ante* profit or loss is close to zero.

follow the market: The strategy of trading on the assumption that the price quotations correctly reflect the market situation.

moment maker: An entrepreneur who deliberately postpones or advances the moment of sale in order to avoid the price risk and/or to make a profit.

monopsony: an exclusive buying monopoly. Nearly all of the EMBs in Tropical Africa were granted a monopsony by their governments. A legal or statutory monopsony did not always entail a *de facto* monopsony.

postharvest interval: Here defined as the interval between the harvest and the shipment of a particular quantity of an export crop.

price risk: The risk of a price fall in the near future. A commodity exporter is exposed to the price risk during the period that he owns stocks, including stocks in sight.

price risk management: Policies, including the judicious use of financial instruments, to reduce (a) the price risk and (b) the period of being exposed to it. Usually used as the equivalent of price risk avoidance.

producer price: The buying price set and publicly announced by a marketing board and held constant during the buying season. The actual buyers are instructed not to pay less than the producer price.

risk-assuming conduct of the exporter: Conduct in which the exporter bears certain risks rather than shifting them to another actor.

risk-avoiding conduct of the exporter: Conduct characterized by a risk-averse attitude. In particular, the price risk is avoided.

selling conduct [of commodity exporter]: The strategies, operations, and behavior of the exporter with regard to export sales.

stocks in sight: Stocks of agricultural products that are expected to be available after the next harvest.

tactical speculation: Selling operations of an exporter involving distant sales which look promising because of a contango premium in the commodity market.

Chapter 8

Commodity TOMCs:
Channel Leaders and Captive Growers

Chapters 8 and 9 discuss internal marketing with special emphasis on the policies of the exporters. (My focus is on policies because the relevant practices have already been discussed.) A key objective of the exporters has nearly always been to gain more control over the upstream actors from whom they bought. Since control, coordination, and management go together, an alternative description of the exporters' policy is that they each wanted to create a partial channel system and aspired to the position of channel leader.

Economists usually focus on one of the exporters' reasons for seeking control: each exporter's desire to keep buying prices down by means of an exclusive position, a monopsony. This desire is an important and permanent factor in internal marketing. However, in Chapter 7, we found another, less well-known explanation for the exporters' desire to gain control: each exporter wants to be sure that he alone receives the product which still is in the hands of the upstream actors. This enables him to engage in early sales and tactical speculation. In this chapter, the word "control" will be used, rather than monopsony. As control is easier in the two-actor pattern, we will discuss this first. Control in the three-actor pattern is dealt with in the next chapter.

GENERAL COMMENTS
ON THE TWO-ACTOR PATTERN

Since the two-actor pattern is inherently unstable (see Chapter 6), we should ask why it survived at all. The main reason for its survival is that in certain sectors it was strongly defended by some exporters. Its most common defenders have been the processor-exporters, with considerable investments in processing installations. Processor-exporters want to safeguard these investments by ensuring captive, regular supplies at reasonable prices. The analysis below primarily applies to sectors dominated by processor-exporters.

When considering the means by which processor-exporters establish control over the growers, our first question must be whether they know the growers by name. During most of the colonial period, this was not the case. The explanation may be sought on the part of the growers (most of whom were illiterate) and on the part of the exporters (who wanted to avoid the administration associated with a large number of individual growers each producing small quantities only). Whatever the reason, the exporters sought a convenient cluster of potential growers, such as the inhabitants of a village. Control could then be effected through the chief and, later, through the officials of a village cooperative. Knowing their names was enough for the exporter. This *indirect control of the growers*[1] worked best in isolated regions, where no alternative cash crops existed and where the people had no tradition of migrant labor. One social drawback of indirect control must be mentioned here. It was tempting for exporters to persuade the chief or other village officials to use force to compel the villagers to cultivate the crop they wanted.

Later in the colonial period, indirect control was replaced by direct control, and individual growers became known by name to the exporters. This corresponds to the contracts known in the West, where the growers enter into a contract with the buyers early in the season, specifying the terms on which growers will later sell to buyers. An exclusive buying right is a normal feature of such forward contracts. Genuine contracts have been introduced in Africa since about 1960 (and have given rise to the literature on *contract farming*, see below), but we should recognize that prototype contracts were used much earlier. Thus, tenant growers on a modern irrigation scheme were generally known by name. Elsewhere, it became customary that an employee of the exporter annually filled out an application form for each participating grower. The form marked an improvement for the latter. Growers could now individually decide whether to continue working with exporters or to withdraw. In the case of annual crops the opportunity to withdraw occurred every year and the exporters soon discovered this form of *countervailing power* of the growers. Exporters learned that it was unwise to ignore signs of dissatisfaction among growers.

One element of the control just described is that it reduced the chance of a rival exporter entering the area where another exporter operated. Competition was eliminated—a feature economists have always been critical of. Economists believed that the absence of competition gave the exporters too much power in the matter of prices. While accepting that this is true, we must recognize that the extent of the exporters' power depended on the circumstances. It was strong when (a) chiefs used compulsion; (b) the region was remote and isolated; and (c) the crop was a perennial one, but it was weak in other circumstances.

THE COTTON SECTOR IN AFRICA

The requirements of international trade are important for the cotton sector. For well over a century, the importers of cotton have not accepted raw or seed cotton, the product that is harvested, but only cotton fiber. (The main reason is that the weight of the fiber is only one-third of the corresponding seed cotton.) It is therefore necessary that the seed is separated from the fiber before shipment. Separation takes place in a simple industrial process called ginning. A further improvement was that the cotton fiber was pressed into bales of standard weight and size, which reduced handling and transport cost, both overland and by sea.

The plant or mill where ginning and baling take place is called a *cotton ginnery*. The *cotton ginner*, that is, the enterprise that owns and/or operates one or more ginneries, is a key actor in this section. Until about 1950, the seed obtained by ginning had no commercial value. A small part of it was used for sowing and the remainder was used as fuel in the ginneries. In the 1950s, a few small oil extracting factories were built near some cotton ginneries. Since then, an increasing portion of the total cotton seed crop in Africa has been processed in this way.

The international requirements and the fact that the cultivation of export cotton in Africa has always taken place on a small scale explain why the cotton growers need one or more complementary actors to do the ginning and exporting. Two situations have occurred in Africa. In the first, which best fits this chapter, there is only one actor: the *ginner-exporter*. In the other situation, which has more similarities with the analysis in the following chapter, there are two separate actors: the *nonexporting ginner* and, farther downstream, the exporter. Both situations will be discussed here because of the opportunity for comparison.

The price which the ginners receive for their cotton fiber depends on the quality of both the seed cotton and the ginning process. The wish to improve the quality of the seed cotton determines the ginners' relations with the growers. To make sure that all cotton is of the same botanical variety most ginners have a system for distributing the right cotton seed to the farmers.[2] They further supply fertilizers and pesticides and supervise cultivation. Having been involved with the crop for several months and having provided farm inputs free of charge (or on credit) gives ginners an exclusive claim on the harvest, even if this is not contractually registered. In cotton channels, a ginner is automatically the channel leader.

Ginning complicates a ginner's storage task because he has to work with three types of stock: seed cotton as input, and cotton fiber and cotton seed as output. When ordering a ginning installation from abroad, the ginner must make a decision on the optimum capacity. As an illustration we may consider two alternative capacity options. One option involves a low-capacity installation: the entire year is needed to process the season's input. The other option involves a fast, high-capacity installation which enables the ginner to complete ginning of the same input within four months. Since this option requires more expensive machinery, average ginning costs are higher. On the other hand the carrying costs, mainly composed of interest charges, although still high, are lower than with the first option because stocks are held for a much shorter period. If interest rates are high, it may be wise for the ginner to invest in a high-capacity ginnery, even if it is idle for eight months out of every year.[3] We now turn from general principles to the individual countries.

Cotton in Particular Countries

Before World War I, cotton was successfully introduced in two African countries. The cotton sector in Nigeria was established by the British Cotton Growing Association which appointed the Niger Company as its buying and ginning agent. In Uganda, the private Uganda Company was the pioneer in erecting a ginnery. However, the real success of cotton came after the war when the colonial authorities began to promote the crop, particularly in remote areas. The French in French Equatorial Africa and the Belgians in the Congo introduced legislation which enabled the authorities to grant cotton *concessions*, i.e., exclusive rights in a certain specified area, to ginner-exporters. The villagers living in a concession were expected to grow cotton and to deliver it to the concession holder. A number of European enterprises were attracted. One of the applicants for concessions was a Dutch trading company which had long operated in the Congo.[4] It was in this period that cotton cultivation started in Chad (Cabot, 1965; Stuerzinger, 1980) and Ubangi-Shari, the most remote colonies of French Equatorial Africa. (As we saw in Chapter 7, the authorities in these colonies had to make the concessions more attractive by the creation of *Caisses de Soutien*.) In Chad, there was a large degree of compulsion, which was abolished only after World War II. Forced labor was also extensively used in the cotton sector in Mozambique.[5]

In the Anglo-Egyptian Sudan, the Gezira Scheme was started in the early 1920s.[6] Technically it was a huge irrigation scheme in the center of the country built and managed by the government, which also allocated land to tenants. Commercially, it was an agricultural project combining one export crop, cotton, with a number of domestic food crops. Because of the cotton, the collaboration of a ginning company was required. The government chose the Sudan Plantations Syndicate (SPS) which was also put in charge of export marketing. From the start, three parties were distinguished: the government, the SPS, and the tenant-farmers. A special feature of the Gezira Scheme was that these three parties received roughly equal shares of the export earnings for cotton, a point to which we will return. For our purposes, the main point is that the SPS was a typical ginner-exporter, which bought directly from the tenant-farmers and had direct control over them. The quality of the Gezira cotton was good and the price high, not only because of its long staple, but also because cultivation, transport, storage, and ginning were carefully coordinated. For many decades, the Scheme was held up as an example for development elsewhere in Africa, including French Africa, where the *Office du Niger* was established in the French Soudan in the 1930s.

Although Uganda was also an important cotton exporter, its cotton sector differed greatly from the one in the Sudan.[7] Since rainfall was sufficient, cultivation could spread to many areas. In the 1920s, many small ginneries were set up all over the country. The majority of them were owned by Indian businessmen. Most of them were nonexporting ginners and depended on another actor to export their output. Thus, a local market for cotton fiber developed. The buyers were either large ginner-exporters (who added the bales of the smaller ginners to their own) or representatives of spinners and merchants abroad.

The following differences existed between the ginner-exporters and the nonexporting ginners. First, while the former were large, the latter tended to be small, usually operating only one ginnery. Second, the ginner-exporters tried to improve the ginning process, while the nonexporting ginners had a high percentage of broken and frayed fiber. Third, it appears that the ginner-exporters tried to keep their ginneries operating throughout the year, while the nonexporting ginners wanted to complete ginning within three or four months. Or, in terms of what we said above, the ginner-exporters preferred a low capacity installation and the nonexporting ginners a high capacity one. This can be satisfactorily explained if we assume that the ginner-exporters could obtain working capital more cheaply. From the perspective of the ginner-exporters it further seemed that there was excess capacity in the ginneries of the nonexporting ginners. In Uganda, this was diagnosed as a recurrent problem in the 1930s and 1940s.

The small scale of many ginneries in Uganda made those ginneries vulnerable to buying by rivals. In the 1930s, the ginners asked for and obtained legislation that created an exclusive zone for each ginnery (not unlike the concessions in French Equatorial Africa, but covering a much smaller area), and restrictions on the transport of seed cotton by road. These restrictive measures were later strongly criticized by Ehrlich (1965) as interference with free competition. Ehrlich's position cannot be assessed in isolation, for he was certainly influenced by Bauer's (1954) attack on the buying monopoly of the new marketing boards in West Africa. In much of anglophone Africa, the criticisms of Bauer and Ehrlich were taken to heart. In the cotton sectors, the old structures were loosened and reorganized in the 1950s and 1960s, mostly as part of a policy to promote cooperatives. In Uganda, Kenya, and Tanganyika, cooperative ginneries replaced private ones to a large extent. Since those ginneries did not engage in export, a new, separate actor was needed in the TOMC: the exporter. In most countries and periods, this exporter was an EMB, but there was also some experimentation with local auctions that enabled the cooperative ginners to sell locally (see Chapter 12).

In francophone Africa, interest in cotton revived after World War II. The *Compagnie Française pour le Développement des Fibres Textiles* (CFDT), a large, semipublic organization, was established in 1948 with the express purpose of introducing cotton growing in French West Africa. In the 1970s, most of the CFDT subsidiaries were replaced by parastatal cotton development companies, but the CFDT remained involved as a shareholder of the parastatals and through technical assistance. At present, it is thus involved in Benin, Burkina Faso, Cameroon, Côte d'Ivoire, Guinea, Madagascar, Mali, Senegal, and Togo—all francophone countries, as well as in Gambia and Guinea Bissau. The extent of its role varies from country to country.

For our purposes, the important point is that the combination of ginning and exporting has been retained in the parastatals that are associated with the CFDT.[8] This arrangement is usually described as vertical integration, i.e., the integration in one organization of all activities between cultivation and export. The performance of the cotton sectors associated with the CFDT seems to be much better than in anglophone Africa (Béroud, 1994). Lele, Van de Walle, and Gbetibouo (1989) also give credit to the vertical integration in the CFDT countries as a factor making for lower costs and higher efficiency. It seems to me that African governments face a dilemma when there is a cotton sector in their country. If they choose the two-actor pattern, there will be a well-coordinated channel system, but the benefits of competition are sacrificed. On the other hand, if African governments choose the three-actor pattern, the growers have more freedom, but it is difficult to coordinate the channel and to achieve high efficiency.

In the 1980s, the parastatals associated with the CFDT began to sponsor village cooperatives. The SODECOTON (Société de Développement du Coton du Cameroun) in Cameroon, for instance, encouraged the formation of village associations and entrusted the primary buying operations to them, paying a fee for each kilogram of seed cotton handled. In this way, the role of the parastatal decreased, while that of the cotton growers increased. Two stages are envisaged in this process. At first, the association is closely supervised by the *moniteur*, the local representative of the SODECOTON. In time, the association qualifies for autonomy and the *moniteur* is withdrawn. The SODECOTON retains the right to reverse the process in case the association no longer functions satisfactorily. In 1993, the majority of the associations were autonomous, but unfortunately, the percentage of associations that lost this status after some time was higher than anticipated.[9] It is too early to say whether the new associations will succeed as actors in the channel, but one point has become clear: the literature on cooperatives in Africa has been one-sided because of its focus on the situation upstream, that is, the relationship with the members.[10] We now see that in some cases, the exporters downstream are the principal beneficiaries of cooperatives and hence eager to establish them and to see them succeed. This is the case when the cooperative helps to secure captive supplies from the farmers in a particular area.

SECOND-PAYMENT SYSTEMS

Under a *second-payment system*, an exporter makes two payments to the grower: the first one when growers deliver their product, and the second one later in the season, when the exporter has sold the product and has been paid by the importer. The price used in the calculation of the first payment is announced beforehand, often at sowing time. At that time, the level of the second payment is still uncertain because it depends on the price which the exporter will finally receive.

The Gezira Scheme was one of the first schemes in Africa where the second-payment system was applied. As we saw, each year's cotton earnings were divided among the tenant-farmers (as a group), the government, and the SPS, along an agreed-upon formula. The farmers received 40 percent; the SPS received 25 percent at first and after 1926, 20 percent; and the government received 35 percent at first, and later, 40 percent. As the tenant-farmers were entitled to their share of the export earnings, the precise amount they got could not be calculated until the end of the season. As tenant-farmers were unable to wait so long, the SPS paid them an advance when they delivered their cotton and a second payment at the end of the season.

It is illuminating to compare the Gezira practice with that of COD at the PMP, which, as we mentioned in Chapter 6, has been long-established and widespread in Tropical Africa. With the COD practice, the transaction was completed when grower and exporter had exchanged produce against cash, but with the Gezira practice, the transaction at the PMP had two phases. It was firmly believed in the Sudan that, on average, growers would receive higher prices under the Gezira practice than under the COD practice.

Since all second-payment systems may be considered to be a variant of the Gezira practice, they are discussed here. We first note that a second-payment system is only realistic in the two-actor pattern. (In the three-actor pattern, there is the complication of the intermediary.) Second, the system requires a great deal of paperwork, including documents supporting a grower's claim to the second payment. Third, the second-payment system represents a re-allocation of marketing functions and risks in the channel. The growers assume more risks, for instance, the price risk, and they provide some seasonal finance because they are prepared to wait longer for the money they are entitled to. As the growers' role expands, their gross income probably increases correspondingly.[11]

While in theory, a second-payment system is superior to the COD system, in practice, the advantages for the growers tend to be meager. One should look at the following aspects. First, the first payment may be low in relation to the total amount received. This is in the exporters' interest because it reduces their seasonal finance requirements. Second, the interval between the first and the second payment may be long, which increases the burden of finance for the growers. Third, the growers may not be sophisticated enough to monitor and assess the exporters' external marketing operations. If the exporters are not competent and/or honest, the grower may not benefit from the system. Trust is an essential element, but it is difficult to retain.[12]

In many countries, the growers have not been enthusiastic when a changeover to a second-payment system was advocated or implemented. This is less surprising when we remember that the growers under the new system provided some of the seasonal finance that was formerly supplied by the exporters. What were the costs to the growers providing this sum? A general answer is hard to give. It depends on (1) access to the banking system; (2) the interest rates charged by formal and informal lenders; and (3) the liquidity position of the rural sector. In countries where the exporters are able to obtain seasonal finance on much better terms than the growers, the net gain of the growers may be small or nonexistent.

Finally, proposals for reorganization at the PMP level should always recognize that growers differ among themselves. Small farmers tend to be more risk averse than large ones. They may thus prefer the COD practice, while large farmers are willing to switch to a second-payment system. Should the interests of one group be subordinated to those of the other? One should further be aware that conditions are rarely constant. Proposals that are ideal for periods with high world prices may be wrong for times of low prices.

CONTRACT FARMING

Under contract farming, the grower cultivates a particular crop in accordance with the terms of a written contract between him and a processor-buyer. The contract is entered into at sowing time and contains clauses about the delivery of the grower's harvest: time and location of delivery, quality standards, possibly a volume limit, and the price. In most cases the buyer is large in relation to the grower. This leads to a contract farming scheme with a large number of parallel contracts. The initiative for a scheme usually comes from the buyer.

If a formal contract is taken as an essential element, contract farming became important in Africa in the 1960s only. From the outset, schemes producing export crops and those initiated by foreign companies have received most attention. There was much enthusiasm because contract farming was associated with modern agribusiness and it diversified the economy; but soon, notably in the 1970s, the negative aspects of these contract schemes were identified. First, the contract created an exclusive relationship and prevented competition, thus depressing the price for the grower. Second, as a rule, the foreign buyer was big and the African farmer was small. In case of a conflict, the grower was no match for the buyer. Third, the buyers had the right to reject produce which they considered substandard. If the buyers abused this right, the grower had no redress. Fourth, the risk of crop failure was wholly borne by the grower.

Now that the first overviews of contract farming in Africa begin to appear (Little and Watts, 1994), a more balanced view is obtained. In exploring the positive aspects of contract farming the vertical relationships between actors in the channel play an important role. Grosh (1993), applying the New Institutional Economics, shows that contract farming is a valuable solution in several situations, for instance, when there is a *first mover* problem. Two actors need each other, but neither is willing to make the first move because the risks are too great. The growers are unwilling to plant a new crop unless they have an assured outlet. Similarly, the processors are unwilling to invest in a factory unless they can count on supplies from growers. Especially if the factory is large, the threshold for a new sector is high. In this situation, contracts help create a viable marketing channel.

Many contract farming schemes have succeeded. The growers' incomes have risen and the number of people applying for a place in the scheme has increased. The growers' position in relationship to the processors tends to improve as the channel is consolidated. However, as with all new initiatives, some schemes have failed, sometimes abruptly, but more often after a period of great uncertainty. The worst features of contract farming seem to have been observed in such periods.

I have inserted this brief discussion of contract farming in the present chapter because cotton farming in the colonial period was in many ways its precursor. Valuable lessons can be learned from studying the free cotton sectors of the past (i.e., sectors without forced labor). I want to make three brief points. First, a remote region offers advantages to both sides. The processor knows that the people have few alternatives and the growers welcome a new opportunity to earn cash, precisely because of the absence of alternatives. Second, an annual crop provides a better negotiating position to the growers because the opportunity to withdraw returns each year. Third, a farming contract may be linked with a tenancy contract in an irrigation or settlement scheme. As this makes withdrawal less attractive for the growers, it strengthens the processor's position.

Glossary

concessions: Here defined as exclusive rights to a raw material in a carefully demarcated area. These concessions were granted by the colonial authorities to a private trader or trader-processor.

contract farming: Farming in accordance with a forward contract between a grower and a buyer.

cotton ginner: The actor who owns and/or operates one or more cotton ginneries.

cotton ginnery: The plant or mill where seed cotton is ginned and baled. In several African countries the cotton ginneries were the first establishments that qualified for classification as industrial enterprise.

countervailing power: Economic power developed by the suppliers or customers of a strong entrepreneur in order to restrain or neutralize the entrepreneur's economic power (Galbraith, 1957). In this book, it refers to the power of growers or intermediaries marshaled to counteract the exporter's power.

first mover problem: A problem which prevents development, for instance, because either the grower or the buyer runs a considerable risk of business failure when he/she makes the first move without knowing whether the other party will respond positively.

ginner-exporter: An entrepreneur who is both a cotton ginner and a cotton exporter.

indirect control over the growers: Exporter's control over growers in a situation where the exporter does not know the growers personally.

nonexporting ginner: A ginner, usually operating on a small scale, who is unable (or unwilling) to engage in exportation.

second-payment system: A marketing system in which the exporter makes two payments to the grower: one when the grower delivers his/her product and another later in the season. While the first payment is based on a provisional price, the second uses the final price minus the provisional one.

Chapter 9

Commodity TOMCs:
Channel Leaders
and Captive Intermediaries

This chapter investigates the ways in which the typical trader-exporter attempts to acquire and maintain control over the actors in the upper section. In doing so, his relationship with the intermediaries from whom he buys is crucial. (If there are several intermediaries, the most downstream one is relevant.) In Chapter 6, we discussed the practices at the IMPs; now, we will turn to the policies at these points, first those of the exporter and then those of the intermediaries. For practical reasons, the analysis in this chapter is confined to West Africa, an area which has been widely recognized as shaped by international trade and small-scale agricultural production.[1]

The periodization in Table 9.1 determines the order of presentation below. We begin with the period before World War I. Since this was largely a period of direct trade, with only a few intermediaries, we can learn little about the relationship between the intermediaries and the exporters. We nevertheless review some elements of this period because these elements help to understand why the exporters had a clear lead in experience after World War I. Next, we turn to the interwar period when the relationship under study was relatively simple: private exporters dealt with private middlemen in a competitive situation. Finally, the postwar period with far-reaching government intervention is reviewed. The relationship between the EMBs, the public exporters, and their *Licensed Buying Agents* (LBA), the intermediaries, reveals interesting parallels with the interwar period.

This chapter derives its fascination from the fact that the middlemen of the interwar period followed policies of their own, upsetting the policies of the exporters. This led to several conflicts, most of

TABLE 9.1. Relationship Between Trader-Exporters and Intermediaries in Commodity TOMCs in West Africa, 1900–1980

Period	Main Designation Intermediary	Main Designation Exporter	Market Structure	Control at IMP	Means of Escaping Control
pre-1914	---	CTC	competition	---	---
interwar	middleman	CTC	competition or pool	tying or market sharing	quality and credit
post-1945	LBA	EMB	monopsony	license	quality

Notes: CTC = colonial trading companies (private enterprise).
EMB = export marketing board (public enterprise): includes organizations in francophone Africa, established in the 1960s, insofar as they physically handled the export product.
LBA = licensed buying agent (mainly private enterprise; cooperatives that operated as an LBA are mentioned in passing). LBA is also used for the *acheteurs agréés* (registered buyers) in francophone Africa.

which have been covered in the literature. In my interpretation of these conflicts, I deviate from conventional views on three points. First, I pay more attention to the periods when the collaboration between the parties was harmonious. Second, I have more sympathy for the intermediaries and approach the conflicts more often from their perspective. (This is the result of my research on the Lebanese traders in Sierra Leone.) Third, I have more economic concepts at my disposal than earlier writers, whose main concepts were monopoly and competition.

THE PERIOD BEFORE WORLD WAR I

The Colonial Trading Companies

The trader-exporters in the West African commodity TOMCs were for the most part the *colonial trading companies*, or "the firms" as they were locally called. The companies were European enterprises in the sense that the directors and shareholders as well as the general managers and part of the staff in West Africa were European. They were also colonial, with British and French companies being predominant.[2] Essentially, they were trading companies which rarely ventured into other activities such as mining, transport, and industry. Most of them were export-import companies because they combined *produce* and *merchandise* trading. (In West Africa, produce referred to the export crops and merchandise referred to imported consumer goods.)

As many companies were large and operated for many decades, we may identify a few by name: UAC (United Africa Company) (Pedler, 1974; Fieldhouse, 1994), SCOA (*Société Commerciale de l'Ouest Africain*) (Coquery-Vidrovitch, 1975), CFAO (*Compagnie Française de l'Afrique Occidentale*) (ibid.), John Holt & Co., and Paterson Zochonis.[3] A number of smaller firms raised the total number to around a dozen. The companies made common cause in trade associations which represented the companies' interests with the colonial authorities. Items that regularly appeared on the agenda were: customs duties, port facilities, railway freight rates, railway carrying capacity, and road building.

Although we will not be studying the colonial trading companies in their entirety, but only as exporters, their merchandise sector cannot be completely disregarded. The merchandise channel and the produce channel were intertwined at the African ports as well as up-country, at least in those areas where produce was grown. The companies therefore spoke of the *two-way trade*. For most of the period under review the farmers used a very high proportion of the receipts for their produce to buy merchandise. (In fact, merchandise often acted as an incentive to grow cash crops.) As the farmers spent the money at once, the two-way trade was highly seasonal.[4] It was essential for the trading companies to estimate the farmers' purchasing power in advance in order to have sufficient merchandise available in the produce areas at the right time.[5] The intertwining also explains why the trading companies were ambivalent about low produce prices. While low prices increased the companies' trading margin, they also reduced sales of merchandise. All this affected the trading companies' relationship with the middlemen, as we shall see below.

The trading companies saw themselves as the leaders of the emerging channel systems. Their claim to channel leadership rested mainly on the innovative role they had played in introducing new commercial practices. Indeed, the trading companies initiated the farmers into the trans-oceanic channel. Their dynamism, another basis for leadership, manifested itself strongly in the early colonial period when the upper sections of the TOMCs for export crops expanded and were modernized. The existing produce drainage systems increased in size, as large areas in the interior of West Africa were opened up through the combined efforts of the colonial governments and the trading companies. The colonial governments constructed modern ports and railways (well-documented investments) and the trading companies established *buying stations*. Each company set out to create its own network of buying stations, guided by its estimates of the agricultural potential of various locations. The extent of these networks varied among companies.

Direct Trade

The main reason why the trading companies invested in these networks of buying stations was to implement a strategy of direct trade. The buying stations provided an outlet for the farmers, most of whom had never before participated in trans-oceanic trade. Obviously, direct trade demanded an extensive network of trading points. That these points were called buying stations is a clear indication that the buying (of produce) was more important than the selling (of merchandise).

The costs of the new buying stations were considerable. This was not only a matter of investment in buildings, but also of staffing. Because of the preference for European staff, there was a European buyer at most buying stations. Some system of supervising the buyers was further necessary and required a number of traveling managers. The whole network also had to be coordinated by the general manager in the port of shipment.

The companies during this time distinguished two types of buying station, the *river stations* and the *rail stations*, according to the evacuation route the companies used.[6] Hence, river companies (with a network of river stations) may be distinguished from rail companies (Van der Laan, 1983a). The initial costs of the river companies were higher because those companies had to invest in a river fleet, such as those on the Niger and the Senegal Rivers. But afterward, river companies were less troubled by competition, as potential rivals were reluctant to invest in a parallel fleet. The situation of the rail companies was the opposite: entry barriers were low and competition was strong. Rail companies enjoyed some valuable advantages over the river companies. For instance, their staff could use the post offices and telegraph services located in the railway towns.[7] The rail companies therefore found internal coordination relatively easy, which strengthened their channel leadership.[8]

In view of the determined efforts made to construct buying stations and to staff and supervise them, I conclude that initially, the trading companies firmly believed in the strategy of direct trade. This belief faded during World War I when staffing became a great problem because many employees returned to Europe to enlist in the army. Direct trade was further undermined by the coming of the lorry and the construction of motor roads in the 1920s.

THE INTERWAR PERIOD

The new motor roads extended the produce drainage systems deeper inland. If the trading companies had extrapolated their prewar policy of advance, they would have established *road stations* along the new roads. However, complacency reigned and the companies did not build such stations—at least not many.[9] The opportunity they missed was seized by a new category of small traders, the middlemen. An Outer Zone developed, which became the preserve of these new middlemen. They bought lorries, often on hire purchase terms (Mars, 1948), and put up simple buildings, which became PMPs on the (company-inspired) principle of "stationary buyer, mobile seller." Since few middlemen could afford a lorry, there was normally only one lorry on each road. The owner carried produce for himself and for other middlemen settled somewhere along the same road.

There were also changes in the Inner Zone. (In Chapter 6, these Zones were defined in terms of primary and intermediate marketing points.) The buying stations of the companies which had been PMPs (with direct trade) became IMPs (with indirect trade).[10] It was at these IMPs that the middlemen and the companies met and traded with each other.

The trading companies were willing to give a free hand to the new middlemen in the Outer Zone, provided their own position as leaders of the entire channel system remained secure. The new middlemen were supposed to operate as a fringe of auxiliary traders backed by the expertise and financial strength of the companies. At the IMPs, the middlemen could sell the produce of the Outer Zone and buy the merchandise wanted by the new produce farmers. Thus, the two-way trade in the Outer Zone was to become a replica and extension of the older two-way trade in the Inner Zone. With the wisdom of hindsight, one must say that the companies were naive in expecting that the middlemen would be content with this auxiliary position, but at the time, this expectation seemed reasonable.

In the early 1920s, it was indeed absurd to think that the middlemen would ever challenge the position of the colonial trading companies. The middlemen were small traders of limited education and sophistication. Their starting capital was negligible and their sphere of action local (the immediate vicinity of the motor road). Nearly everywhere, they had to reckon with potential rivals because many others were interested in trading and the entry barriers were low. On the other hand, the companies were well-established and experienced. They were few in number, but large and financially strong. Moreover, they had a stronghold in the export trade, where entry barriers had become formidable. Although the odds were against them, the middlemen grew in strength during the interwar period. We must now examine this process.

Conditions at the Intermediate Marketing Points

Trade at the IMP took place between the *factors*,[11] the employees in charge of a trading company's buying station, and the middlemen. We must first ask how many factors and middlemen there were, and in particular, how many companies operated at an IMP. At most IMPs, there were several companies—a situation to be discussed in detail below. Before doing so, we must briefly look at the situation of "one-company-at-one-IMP," which was more common than is often assumed. The situation occurred in areas where the total volume of produce was too low to encourage the coming of a second company. It also occurred along the navigable rivers where river companies often had a *de facto* monopoly. This, incidentally, is a reason for me to say little about the river IMPs in this chapter and to consider primarily the rail IMPs. Finally, while the number of companies at an IMP normally increased over time, the opposite happened when companies merged. From the local perspective, a merger was desirable for the channel leader, particularly if it meant that only one company remained at the IMP.

The geographical conditions at the IMPs were also important. Nearly all the new roads were so-called *feeder roads*, which fed the produce-carrying lorries to the railway or, less often, a navigable river. These roads were not linked to others, but began at a railway station and ended somewhere at a village in "the bush." Since a lorry operating on such a road could not travel any farther, the produce had to be unloaded at the IMPs near the railway stations. The break in transport and the attendant necessity to handle the produce made the IMPs a natural point for trade. It was practically impossible for a middleman to bypass the companies at the IMPs. We note that the same held true for the merchandise channel. In theory, it was possible for a middleman to travel by train to the port and to buy directly from importers, but in practice, he bought from a company at the IMP. Thus, there was a double dependence and the middlemen's wish to escape from it was stimulated on both the produce side and the merchandise side.

It is plausible that the trading companies wanted to preserve the system of isolated feeder roads. In this, they were the natural allies of the railways whose general managers normally objected to the construction of a unified road system. In spite of opposition, new types of road were built in the late 1920s, including some linking feeder roads with each other. (Colonial officials often pressed for these roads because it saved them traveling time.) As a result, middlemen could send their lorries to more IMPs than before. Middlemen preferred IMPs closer to the port of shipment where prices for produce were normally higher, and IMPs where several companies competed.

Competition at the IMPs and the Tying System

When two or more companies operated at an IMP, a middleman could hope to mitigate his dependent position by playing one company factor off against another—a special, covert form of competition. The trading companies were scared of this and tried to prevent it in several ways, such as market sharing agreements and "tying." Much has been written about the market sharing agreements (which were countrywide) and little about tying, which had only local effects.[12] But from the MCA perspective the latter is highly interesting. Under the *tying system,* each company tried to tie a number of middlemen to itself in an exclusive relationship covering both produce and merchandise transactions. At each IMP, each company attempted to ensure it had one tied middleman (or more if there were several PMPs along the motor road). It was the factor who selected the right middlemen and initiated and maintained the tying arrangement. Thus, competition was circumvented and channel control retained.

Of course, the middlemen realized that, if prevented from dealing with a rival company, they forfeited the benefits of competition. Middlemen were willing to accept this drawback because of credit offered by the tying companies.[13] Indeed, the companies used trade credit rather than contracts to keep the middlemen in a dependent position. Credit was used in two ways. First, a company sold merchandise on credit so that the middlemen did not have to finance their own stocks.[14] Second, the companies gave seasonal advances so that the middlemen could buy a larger volume of export crops than their own resources permitted. Tied middlemen experienced an annual debt cycle with a peak at harvest time, declining to a zero position soon afterward. It was this debt—the more the better—that enabled middlemen to increase their turnover.

The tying system turned free traders into captive middlemen. This gave the companies a commercial advantage. It allowed them to pay lower buying prices and thus to earn a higher trading margin. Moreover, although causing rigidity in the channel, the tying arrangement left the factor great freedom in the competition with rival companies. Normally, the volume of produce passing through an IMP was fixed and could not increase in the short run. If a factor wanted a larger share of this volume, he had to get hold of it at the expense of a rival company. If a factor was successful in getting the right middlemen and if he gave them appropriate help, he could raise his share of local trade. In sum, the earlier changeover to indirect trade had weakened the channel leader's control (compared to direct trade), but by having tied or captive middlemen the channel leader was able to reestablish control over supplies.

The costs of the tying system were considerable for the companies. A great deal of their capital was locked up for many months and earned no interest, as it was used as trade credit. As a rule, these costs were difficult to measure because credit injected into the produce channel could not be separated from that injected into the merchandise channel. In fact, the dovetailing of the two credit flows may well have offered a great challenge to those managing the company's cash flows. As long as the outstanding trade credit fell to zero once a year, the system was sound in spite of the middlemen's inability to offer collateral. However, problems occurred when the middlemen could not repay in time. These problems manifested themselves during the crisis of the 1930s.

The companies were, in fact, the key figures in the allocation of seasonal finance in the interwar period. They made credit available to their tied middlemen, obtaining, as we saw, a reward in the form of favorable prices rather than of interest payments. In addition to using their own capital, the companies borrowed from the commercial banks on a seasonal basis. As few middlemen were in a position to borrow directly from the banks, the banking system could remain small and geographically confined to the ports. Put differently, the trading companies were the principal "bankers" of the two-way trade.

Factors and Middlemen

The actual decisions to make the tying system work were made at the IMPs. There, the middleman regularly met a company's factor. The factor was in a position to assess the character and the past performance of a particular middleman. How much credit could be safely extended? How could loyalty be secured? Was there a rival company who wanted to take the middleman on? These were the questions considered in normal and prosperous years. In times of recession, as in the 1930s, the nature of the questions changed and the worries of the factor increased. If there were arrears in credit repayment, was this a mere accident? Was the solvency of the middleman at stake? If so, was it better to drop him and write these advances off? Or would it be better to continue for another year, hoping that there would be a turn for the better in the coming season? After all, the recession would not last forever!

While a company's factor was not personally liable for default by a middleman, the need to write off bad debts did not improve his career prospects. On the other hand, if a factor was too restrictive with credit, his turnover would decline, which also endangered his career. The worries and anxieties of the factors should therefore not be underrated. Nor should the incentive for a middleman to keep some of his operations secret, notably when he was in arrears. If a middleman could not get merchandise on credit from his own company, he would secretly buy from another company on a cash basis (with the last of his money!), rather than sit in an empty shop. Obviously, a recession greatly endangered the tying system, quite apart from the structural problems, to which we must now turn.

One structural problem of the tying system resulted from the fact that credit had to be repaid in produce. A company factor therefore welcomed a tied middleman when he arrived with a lorry full of produce, but if the quality turned out to be poor, the factor faced a dilemma. If he rejected the produce, he would have to wait longer for repayment; if the factor accepted, he endangered the export quality and the reputation of his company abroad. How tough should a company factor be when examining quality? It was this dilemma that made the companies great protagonists of produce inspection by government officials. When studying the arguments preceding the introduction of *produce inspection services* in British West Africa in the 1920s, one notices two proposals: inspection just before shipment (to protect the reputation of the country—fully endorsed by the governors) and inspection up-country. The companies had little reason to support inspection before shipment because it tended to make their own carefully built-up international reputation superfluous. However, inspection up-country protected companies against shrewd or careless middlemen; and a company's factor could appreciate the fact that the frustration of a middleman whose produce was rejected would be directed at an inspector and not at the factor.

The trading companies thus had three complaints against the middlemen. First, middlemen did not fully repay the credit extended to them. Second, middlemen evaded their tying obligations and were disloyal and ungrateful to the company that had generously helped them with finance. Third, many middlemen delivered produce of poor quality—poor, but not bad enough to be rejected.

I see a connection between these complaints and the deterioration of the middlemen's public image during the 1930s. This deterioration was very clear for foreign middlemen such as the Lebanese. True, the Lebanese had been in disfavor with some of the African people in the coastal cities (e.g., the Creoles in Freetown), but until 1930, they had been defended by many Europeans. All this changed in the 1930s as a result of the changing attitude within the trading companies. Why the middlemen began to have a negative image is less clear in countries such as the Gold Coast and Nigeria where the middlemen were mainly Africans. Even so, in those countries, the middlemen were also being criticized in the late 1930s (see below).[15]

As the tying system eroded and the relationship between factors and middlemen soured in the 1930s, the trading companies were more inclined to resort again to agreements among themselves to restrain competition, fix common buying prices, and freeze market shares. Fixed market shares were the ideal method to counteract the attempts of middlemen to play one company factor off against another. In this connection, it is important to remember that the economic crisis of the early 1930s forced the companies to repatriate part of their European staff and to replace those people with cheaper African employees. As a rule, these men had less freedom in their negotiations with the middlemen, which weakened the position of the companies at the IMPs. This increased the companies' desire to come to market-sharing agreements. However, the middlemen had another arrow to their bow.

Moment Making in the Cocoa TOMCs

Essentially, what is recounted here is a new stratagem of the middlemen and the subsequent attempt of the companies to neutralize it. The scene was the cocoa sector in the Gold Coast (and to a lesser extent, Nigeria). The time was the 1930s. In my account, the story begins with the middlemen whose initiative was the source of serious problems, but I must warn the reader that in practically all other accounts (Nowell Report, 1938; Milburn, 1970; Ehrler, 1977; Meredith, 1988; Fieldhouse, 1994), the companies are portrayed as the actors who upset the status quo.

According to the general accounts, events were set in motion by the Cocoa Buying Agreement (or Cocoa Pool as it was popularly called) which the companies concluded in September 1937. The Cocoa Pool led to dramatic protests in West Africa, which culminated in a highly effective *cocoa holdup* in the Gold Coast and a (less effective) boycott of merchandise. This dramatic turn of affairs surprised both the companies and the colonial authorities and led to the establishment of a commission of inquiry, the Nowell Commission. The commission spent three months in the Gold Coast and Nigeria (where there was a milder form of protest) and finally recommended that the companies dissolve their agreement. This was done and the cocoa flow resumed in May 1938. There is a general consensus about the historic significance of the conflict, but not about the proper interpretation. Some see the companies arrayed against the farmers, while others, including myself, think of the middlemen as the opponents of the companies.[16]

When the companies defended themselves at the hearings of the Nowell Commission, they complained about the middlemen. A crucial complaint concerned the middlemen's speculative behavior, which expressed itself in the shrewd timing of their cocoa deliveries to the buying stations of the trading companies. For many years, timing had been a mechanical decision. As soon as a middleman had accumulated enough cocoa, he transported it to a buying station; this changed in the 1930s. The companies noticed that deliveries became irregular. When world prices were rising and the companies were about to raise their buying prices in response, the middlemen held back cocoa. Similarly, when world prices were falling, the middlemen delivered as much as possible. This judicious timing, this moment making, increased their profits.[17] Apparently, some middlemen knew more about world prices than in the past. However, the source of their knowledge was not investigated. Perhaps middlemen obtained news from friends among the company employees. In any case, it was unfortunate for the companies that this happened when they themselves were hard put to survive.

The bitterness of the companies bordered on indignation. Many company employees felt that the middlemen's behavior was illegal and a form of cheating. To explain this view we must consider the cocoa stocks in possession of the middlemen. Who was the owner of these stocks? The answer depended on the legal position of the particular middleman. If he was an agent, the stocks belonged to the company, but if he was an independent trader, the stocks were his own. Logistically, this difference did not matter, but in case of stock appreciation or depreciation as a result of world price changes, it became extremely important. A middleman who was merely an agent had no right to benefit from stock appreciation since the stocks were owned by his principal. Agents were therefore often asked by their principals to declare the stocks in their possession. In the 1930s, the companies discovered that their agents frequently made false declarations. When an agent was asked to make a declaration, he had good reason to expect that a price change was in the air. If an agent expected the new price to be higher, it was in his interest to "underdeclare." When delivering to the company afterward, he was paid the old price for the quantity declared and the new price for the undisclosed quantity. An agent thus pocketed part of the appreciation of the stock in his possession, which legally belonged to the company. Similarly, if an agent expected a price fall, he would "overdeclare" (Van der Laan, 1987b).

One element of the Cocoa Buying Agreement of 1937 was a genuine innovation designed to neutralize the moment-making practices of the middlemen.[18] To implement this innovation, the companies agreed that the common buying price would be fixed (and changed) in London by a top employee of one of them and the price be cabled to Accra (and Lagos) and then relayed to the companies' buying stations. Speed in communicating price changes was an essential element. The companies wanted to link their buying prices as closely as possible, timewise, to world prices. With a relay interval of at most 48 hours—too short for the middlemen to advance the moment of delivery—they expected to foil the middlemen's speculative behavior. Because of the early dissolution of the agreement, the innovation was never properly tested.[19]

What was the effect of this episode on the companies? I gather that they were disillusioned. They had come up with an innovation, but it had been swept away by the authorities. Also, it was now patently clear that African opposition could not be ignored. Would it have taken the companies long to come up with a new strategy? We do not know because the war intervened in September 1939 and led to statutory marketing by the British government.

Interpreting the Interwar Changes

How do we interpret the interwar period? Why did the strength of the middlemen vis-à-vis the trading companies increase so much? As far as I can see, this question has never been asked explicitly. This was partly because the new power of the intermediaries was a great surprise, totally unexpected around 1920; it was partly because the exigencies of World War II prevented early systematic reflection; and it was partly because the trading companies had their own subjective answer: by their dishonesty, the middlemen had got the better of the honest trading companies. The companies did not have to be ashamed of their defeat; who could win against crooks? To my mind, this view, with its moral overtones, has had greater influence in West Africa than has been recognized in the academic literature.

I favor another explanation—a dynamic one—which takes into account changes that occurred over time. The success of the middlemen was the result of gradually acquired skills and experience. Their knowledge, notably of grading and world prices, had increased so much that in some transactions, middlemen were a match for the exporters. Some years ago, I described this as countervailing power (Van der Laan, 1983a). Now I prefer an explanation, which is not essentially different, but is couched in marketing channel terms. Although downstream actors had an initial economic advantage over upstream actors, this advantage eroded over time as the upstream actors gained experience and devised their own commercial strategies. Applied to West Africa, I hold that there was a successful process of commercial emancipation among the middlemen during the interwar period.[20] As a result, the trading companies' share of the total trading margin fell, while the middlemen's share increased.

It is a pity that the hypothesis of commercial emancipation was not formulated in the 1930s. It might have inspired the authorities to look for a similar process of emancipation among the farmers, notably in areas where produce had been cultivated for a long time. It is quite plausible that farmers had become more knowledgeable about grading, prices, and gains to be had from competition among the middlemen. Instead of investigating the possibility that farmers were becoming better equipped to defend their own interests, the authorities stuck to the view that they needed permanent protection against traders. This greatly influenced the actual postwar policies (Ehrlich, 1973). Worries about the dishonesty of the middlemen led to elaborate legislation, which inevitably caused rigidity in internal marketing. In short, the mistaken interpretation of the middleman's success was responsible for mistaken policies after the war.

THE PERIOD AFTER WORLD WAR II

We recall that public exporters were established in many African countries in the period after World War II and remained dominant until the structural adjustment policies of the 1980s. To allow for a proper comparison of the relationship between exporters and intermediaries for both the interwar and the postwar periods, we consider here only those public exporters that were engaged in the physical handling of produce. Below, these public exporters are designated as EMBs and the intermediaries as Licensed Buying Agents (LBA).[21] Our focus is on the transactions at the IMPs where the LBAs sold lorry loads of produce to the EMBs.

The predominant approach in the literature on the EMBs starts with the theory of monopoly. Obviously, an EMB was in an exceptionally strong position because of the legal monopsony granted by the government. Not only did the farmers have to contend with this strong position—a point extensively discussed in the literature—but the LBAs did too: the prewar strategy of playing one exporter off against another was no longer possible. The same conclusions are reached by MCA. Since there was now only one channel system, there was no possibility for the intermediary to escape from a channel leader's control. He was a truly "captive" intermediary and where smuggling was impossible, his dependent position was more pronounced than in the interwar period. In fact, it may be argued that the EMBs were established precisely to hold the middlemen captive or, as

Deutsch (1990) puts it, to educate or discipline them. So much for the theory. We will now turn to the actual situation.

New Channel Leaders and Their Means of Control

The new EMBs were powerful channel leaders who had three means of control at their disposal: finance, contracts, and licenses. The last one, backed up by inspection, became the chief instrument of control, both in anglophone and francophone Africa. The force of this instrument had three bases: (1) a restrictive policy was applied in granting licenses; (2) renewal, often required for each new season, was not automatic; and (3) revocation of the license in case of unsatisfactory performance. In most countries, the number of applicants for a license was high, which helped to give the EMB effective control over the LBAs.

However, licensing had two weaknesses. First, the authority to license was given to a composite body, in which other interests besides those of the EMB were represented. This blurred and weakened the EMBs attempts to control its LBAs. Second, licenses are a negative instrument (in contrast to contracts which encourage the actors to build up a good reputation). Licenses created or reinforced a pervasive climate of distrust. Distrust of middlemen, we recall, had marked the attitude of the companies in the 1930s and part of it carried over into the postwar period, especially as many LBAs had been middlemen before the war.

In view of the licensing system and its significance, the word "agent" in anglophone Africa was a misnomer. It suggested a relationship of trust between the EMB as principal and the LBA as agent. It also suggested a reliance on the law of contract rather than enforcement by legislation and a licensing system. In fact, the term "Licensed Buying Agent" was highly ambiguous. Was the intermediary an independent trader, restricted by the conditions of a public license, or was he a mere agent, dependent on instructions from his principal?[22] This ambiguity was a cause of misunderstandings and irritations (see below). It is significant that in francophone Africa, where public intervention came later, the terminology was much clearer. An *acheteur agréé* (registered buyer) was an independent trader.

Why then was the word "agent" chosen and retained in anglophone Africa? One reason was historical. In the early years, the

EMBs in British Africa left the physical handling of the commodities to the colonial trading companies whom the EMBs appointed as their buying and shipping agents. As soon as the EMBs began to handle the crops themselves, they should have stopped using the word "agent." Unfortunately, they did not. The new cohort of LBAs, appointed in the 1950s and 1960s when the trading companies withdrew from the produce sector,[23] were incorrectly called agents. The second reason was a corollary of the authorities' view regarding the income of the LBAs. In the 1940s and 1950s, the authorities strongly felt that the LBA income should not include a trading profit, but be merely a reward in the form of a commission. Thus, the early EMBs fixed rates of commission per ton of product. Commissions, obviously, were earned by agents.

The EMBs did not use credit as an instrument of control to the same extent as the companies had done. This is surprising because the LBAs were just as eager as the middlemen of the interwar period to obtain seasonal finance from the exporter. Moreover, because of their public status, the EMBs had easy access to bank credit—easier than the trading companies before them.[24] There certainly was a substantial financial flow from EMBs to intermediaries after World War II. Why then was credit not as powerful an instrument of control as before?[25]

In Africa, obtaining finance has been mainly viewed as a privilege. It was, for instance, the privilege of the EMBs to receive either credit from the central bank or a treasury guarantee that enabled them to borrow from the commercial banks. It was also the privilege of the LBAs to receive advances from the exporter. It was in keeping with this perspective that in many countries, the EMBs made the amount and the terms of their advances dependent on the legal status of the LBA. More precisely, they were more generous toward cooperative than private LBAs. Among the latter, EMBs helped national rather than foreign LBAs. A political agenda thus crept in: the EMB ought to help and stimulate the cooperatives and, if private traders were also to be helped, nationals deserved more assistance.

Clearly, this perspective was not compatible with the idea of using seasonal credit as an instrument of control and discipline. The consequences of the privilege idea were predictable. Private LBAs searched for other sources of finance to protect themselves against the risk that the EMB might reduce or withdraw financial facilities. Insofar as they were successful, the EMBs' financial instrument became a blunted weapon. Cooperatives did not fear being disciplined by the EMBs because any attempt in this direction would be met with rousing public opinion against the hostile act perpetrated against the cooperatives and the farmers! Many EMBs learned that it was politically unwise to criticize or oust a cooperative LBA.

Because of the absence of effective sanctions, the relationship between an EMB and a cooperative LBA could become very tense.[26] This may come as a surprise because it is often argued that the relationship would perforce be harmonious, as neither party was driven by the profit motive. It has also been frequently asserted that the early EMBs acted as an umbrella for the cooperatives by removing the commercial risk (Ord and Livingstone, 1969). Indeed, the cooperative movement grew rapidly in rural Africa in the 1950s. However, another element needs to be considered. Both parties were motivated by the desire to avoid losses, and this was often as powerful a source of conflict as the profit motive of private traders!

Attempts to Escape the EMBs' Control

The LBAs attempted to escape the EMBs' control and/or to seek a profit on top of their commission payments in several ways. The most controversial attempts centered on the issue of quality; they occurred with those commodities where the EMB distinguished two or more grades, each with its own producer price and commission. It must be pointed out that the early EMBs in anglophone Africa were eager to improve the average export quality and, hoping to achieve this, had introduced several grades and raised the producer price for the top grade(s).[27] This was a policy which the private trading companies had never adopted in the interwar period in spite of pleas by the authorities to do so. The EMBs' policy quickly proved successful. The quality of cocoa exported from the Gold Coast/Ghana and Nigeria improved significantly in the 1950s (Ord and Livingstone, 1969).

The EMBs expected the LBAs to be neutral (and disinterested) when bulking the parcels bought at the PMPs. Thus, parcels classified as Grade I (the highest quality grade) at the PMP should have formed a Grade I flow which the LBA should have kept separate from the flow consisting of Grade II parcels. (This would have been part of a good agent's job.) In fact, the LBAs mixed the flows: they added some Grade II parcels to the Grade I flow, thereby diluting the average quality of the top grade. On the parcels thus transferred, LBAs made a profit equivalent to the margin between the producer prices for the two grades. Mixing demanded skill because the quality of the Grade I loads delivered to the EMB had to be just above the threshold specified in the grading regulations.

The EMBs saw the volume of their Grade I exports rise, but were disappointed that the average quality declined. In those cases where total earnings on Grade I exports decreased, the EMBs bitterly complained about the LBAs whose "greedy" *mixing practices* reduced EMB earnings. Two remarks are in order. First, if the LBA was an independent trader and thus the true owner of the commodity in his possession, he was free to bulk as he saw fit. After all, he bore the physical risks of stockholding, including that of quality deterioration and inaccurate examination before purchase. In this light, I see no reason to condemn mixing as being either illegal or immoral. On the other hand, if the LBA was an agent, he could be rightly accused of cheating his principal. Unless we are clear about an LBA's legal status, we have no proper standard for judging his behavior. Second, the demarcation of grades and the setting of producer prices for these grades are delicate operations.[28] This second point is also borne out by the experience of the *Caisses de Stabilisation* in francophone Africa. The *caisses* found that their licensed exporters engaged in similar mixing practices.[29]

Mixing was not the only practice that led to irritation and exasperation among the EMBs. There were also many allegations that some LBAs falsified the way bills so as to claim a higher transport allowance.[30] Here, the clearly fraudulent practice was to indicate a PMP that was farther away from an EMB's receiving depot than the actual PMP where the lorry load came from. Other elements of the transactions at the IMP (weighing, new or used bags, speed of payment, document procedures, etc.) required both EMBs and LBAs to be constantly alert and could easily cause frictions. On the other hand, in some periods, the LBAs complained that the level of the commission was too low. We must guard against explanations that are purely phrased in terms of human faults. Instead, we should remember that when a government rules out price competition by means of administered prices, the competitive struggle tends to reemerge in the secondary aspects of the transaction. This, I believe, happened at the IMPs.

The irritation just mentioned was a factor in the seesaw development of the EMBs: the 1970s were a period of expansion and the 1980s one of contraction (normally as part of structural adjustment policies). One result of this seesaw process was the variation in the number of places up-country where an EMB operated a depot and took delivery of produce. Theoretically, we may distinguish two types of EMB: the "light" ones, whose physical presence is confined to the port of shipment, and the "heavy" or vertically integrated ones, with a (larger or smaller) network of depots up-country. In the early years, it was felt that a light organization sufficed, as long as it operated from the "commanding heights" in the port of shipment. From there, a light EMB could control the whole channel system. Later, notably in the 1970s, many politicians felt that some degree of vertical integration was desirable, even if it meant more employees, buildings, and lorries. This was implemented in many countries.

Vertical integration eliminated the LBAs and replaced them with internal staff. The expectation was that control over internal staff would be more effective and that the problems caused by obstinate or dishonest LBAs would disappear.[31] In fact, vertical integration was a return to direct trade! I mention the issue of control because I believe that it counted for more in the 1970s than the issue of costs. True, some officials expected that a vertically integrated EMB could operate more cheaply than one employing LBAs, but it seems unlikely, even in the 1970s, that this view went completely unchallenged. Similarly, I believe that the EMBs were willing to accept contraction in the 1980s because vertical integration had disappointed them on both scores. First, instead of falling off, marketing costs had often grown more than proportionally. Hence, a return to LBAs seemed to secure cost reductions in favor of the farmers. Second, control had proved less effective than had been hoped in the 1970s. *Noncompliance* by LBAs had indeed been eliminated, but new problems had arisen because of noncompliance by the EMBs' own staff.[32] In short, control by the channel leader over operations up-country by means of vertical integration had proved to be elusive.

Glossary

buying station: A building at which the employees of a colonial trading company used to buy produce and sell merchandise.

cocoa holdup: A rural protest movement, manifesting itself in the refusal of cocoa farmers to sell their crop.

colonial trading companies: Foreign trading companies operating in the colonial period which engaged in exporting produce and importing merchandise. They were the exporters of the three-actor pattern.

factor: Employee in charge of a buying station.

feeder road: A road built to feed produce to a railway station or a river wharf. The construction of such feeder roads was given a high priority in colonial territories where the railways operated below capacity.

Licensed Buying Agent (LBA): A private trader or cooperative society appointed and licensed by an EMB. An LBA must comply with the EMB's instructions.

merchandise [as used in West Africa]**:** A collective name for imported goods, in particular, those for mass consumption.

mixing practices [in internal marketing]**:** An intermediary's practice of combining and mixing lower quality lots of a particular crop with those of higher quality.

noncompliance: Failure of employees, agents, or licensees to comply with instructions given by the employer, the principal, or the licensing authority. The latter often try to reduce the incidence of noncompliance by disciplinary measures.

produce [as used in West Africa]: A collective name for all agricultural export products grown by small-scale farmers.

produce inspection services: Government department whose inspectors examine and grade export crops. Inspection takes place in the port of shipment, at many IMPs, and, in some countries, also at the PMPs.

rail station [of a colonial trading company]: A buying station that depended on rail transport. A colonial trading company that operated such rail stations may be described as a rail company.

river station [of a colonial trading company]: A buying station that depended on river transport. A colonial trading company that operated such river stations may be described as a river company.

road station [of a colonial trading company]: A buying station that depended on road transport.

tying system: An exclusive relationship between an intermediary and his exporter covering both produce and merchandise transactions.

two-way trade [as used in West Africa]: A trading system in which the collecting produce trade and the distributing merchandise trade are strongly intertwined.

Chapter 10

Commodity TOMCs:
The Goal of Rapid
Produce Evacuation

The term *rapid produce evacuation* was frequently used by the European exporters who operated in Africa before 1940. Evacuation refers to the removal of the current export crop from the country of production. Since economists have paid little attention to the motives for, and the consequences of, rapid evacuation, such an exploration is attempted here. It involves four points. First, the notion of rapid evacuation is reconstructed. It is then explained in terms of actor decisions, particularly in the fields of financing and risk bearing. Third, the macroeconomic consequences, both in Africa and in the countries of destination, are reviewed. Fourth, rapid evacuation is assessed: Is it beneficial for the producers and the African economies?

THE NOTION OF RAPID EVACUATION

The picture below is reconstructed from various sources. Since most of the written remarks refer to the situation in West Africa during the first four decades of this century, the special features of this region at that time should be briefly recalled: (a) all exporters were private businesspeople; (b) nearly all of them were trader-exporters and bought from middlemen; and (c) the main export crops were characterized by a definite harvesting season.

Evacuation from the country took place in two stages. In the *first stage of evacuation,* the crop was transported from the PMPs up-country to the warehouses in the port of shipment. In the *second stage of evacuation*, consignments were prepared and then loaded as soon as a ship for the right destination was available. Often, the term "evacuation" referred to the first stage only, partly because this first part was seen as the most challenging component of the evacuation program. As we saw in Chapter 6, temporary storage was necessary at the PMPs and the IMPs because the product had to be transferred from one mode of transport to another. Exporters and intermediaries saw this *intermezzo storage* as an unwanted element in the program and tried to keep it as brief as possible in order to raise the average rate of evacuation for the entire first stage.

Some connotations of rapid evacuation were similar to those associated with the wartime evacuation of people. First, the evacuees would be transported from a dangerous zone to a safe one. In the same way, the crop was seen as exposed to various risks as long as it was in Africa. Second, as the lives of the evacuees are at stake, the highest possible speed is justified and all means of transport should be used. In the same way, the logistic goal of rapid evacuation of the crop had an extremely high priority.

Some aspects of rapid evacuation were easy to observe and document. First, the means of transport were used to capacity during the evacuation season. In countries where the railways played a large part in the evacuation to the port of shipment, their role is well documented—a welcome source of information on rapid evacuation. Second, the evacuation season was a hectic time for traders and transporters. The permanent staff worked long hours and a great deal of casual labor was temporarily employed.

Did rapid evacuation become a goal in itself? There are a number of indications in this direction. First, there was a sense of satisfaction when the ship that carried the last consignment of a year's crop left earlier than in previous years. It was almost seen as a game (which suggested the metaphor of the relay run to me). Similarly, there was a sense of failure when evacuation was completed late. (Even if the explanation was convincing—a bumper crop, a broken railway bridge, or heavy rainfall—it was felt as a defeat.) Such sentiments suggest a narrow view of the matter. Second, exporters were upset when the intermezzo storage took longer than normal due to insufficient transport capacity. The close relationship between storage capacity and transport capacity seemed to be over-looked.[1] Third, no worries were expressed about the costs of idle resources during the off-season. Indeed, the means of transport, the buildings, and the permanent staff were largely idle in this period. However, nobody seems to have suggested that with a slower rate of evacuation, fewer resources would be needed. In fact, people spoke positively about the off-season. It was seen as a time to finally relax, to catch up on arrears, to clean and repair the produce stores, and to go on leave.

Governments were not directly involved in rapid evacuation, but when the exporters complained about the infrastructure (the capacity of the railways, the port facilities, etc.), the authorities had to say how high a priority rapid evacuation had in their eyes. Although more aware of the costs of having idle resources in the off-season than businesspeople, governments do not seem to have expressed doubts about the goal of rapid evacuation.

This picture seemed strange to me. In the literature on optimal stockholding strategies, interest rates and price expectations are recognized as crucial variables. (High interest rates are an incentive to sell one's stocks, but the expectation of rising prices is a reason to hold on to them.) However, in pre-1940 West Africa, changes in interest rates and rising prices were rarely mentioned. Did business-people pursue rapid evacuation regardless of these factors? But that seemed irrational from an economic point of view! This enigma forced me to look more closely at the conditions in the upper section of the TOMC, particularly with regard to risks and financial requirements. I then found that rapid evacuation was not at all an irrational phenomenon, as we shall now see.

ACTOR DECISIONS PROMPTING
RAPID EVACUATION

Since stocks are a crucial variable in the notion of rapid evacuation, we should first look at the *stock owners*. Among them, we identify the farmers, the intermediaries, and the trader-exporters in Africa (in the three-actor pattern), as well as the importers abroad. Farmers and intermediaries meet at the PMPs and the latter meet the exporters at the IMPs. As sale and stock delivery coincide at the

PMPs and the IMPs, there is no need to distinguish economic from physical stocks at these points. The situation is more complicated at the export level (see below).

In Table 10.1 the cost elements of stockholding are summarized. Since for some items neither the actual nor the opportunity costs can be measured, I needed a new term. I chose *burden* because it has no pretention to precision. It is a special feature of Africa that the burdens are of several kinds and large enough to influence decision making. (Although costs and burdens cannot be added up, *total burden* covers them both.) It may seem that the word burden is not needed in the economies of the West because all the carrying costs of stockholding can be measured. Indeed, even the elusive price risk can be approximately measured, at least if the owner has hedged his risk in the terminal market. However, I hold that an uncovered price risk is a burden, even in the West.

In Chapter 7, we encountered the term "long position" and the significance of the judicious timing of purchases and sales. We also saw that the high marketability of the commodities makes timing flexible: one can find a buyer at almost any time. One can therefore reduce the volume of one's stocks immediately. Since these elements apply to exporters and intermediaries, they want to have accurate and up-to-date information about their stocks. For them, "stock at the end of the day" is an important concept.

TABLE 10.1. Costs and Burdens of Stockholding in the Three-Actor Upper Section of a Commodity TOMC

	exporter	intermediary	farmer
1. Finance, external (costs)	x	xx*	--
2. Finance, internal (burden)	x	xx	xxx
3. Physical risks, insured (costs)	x	x*	--
4. Physical risks, not insured (burden)	x	xx	xxx
5. Storage, hired store (costs)	x	--	--
6. Storage, own store (burden)	x	x	x
7. Price risk (burden)	xxx	xx	x

xxx > xx > x
*   = available to bigger intermediaries only
-- = absent

Insofar as sale and delivery coincide, stocks and the related total burden move simultaneously downward in the channel. For example, the total burden of an intermediary begins when he buys and takes delivery of the produce; it comes to an end when he sells. It is clear that the intermediary can reduce his total burden by shortening the interval between the purchase and sale of a particular quantity. The total burden acts as an incentive to speed up the product flow. In some cases, as the table indicates, the total burden per ton is higher upstream than downstream, strengthening the upstream actors' desire to sell. It is now time to review the various cost and burden elements.

Finance: Costs and Burdens

An exporter needs a large amount of working capital. Most of it corresponds to the volume of his *preshipment stocks*. The typical exporter has a reasonably accurate idea of his *seasonal finance requirements* before he starts buying. An exporter generally has two sources of finance to meet his needs: his own capital and external funds, notably from commercial banks. The costs of bank credit are easy to measure because they are equal to the interest and other costs charged by the bank. (Insofar as an exporter uses his own capital, bank charges are the best approximation of his costs.) It is unattractive for an exporter to employ exclusively his own capital because it will then be idle during the off-season. A preferred strategy is to use large sums of seasonal bank credit.

Since the financial aspect tends to be underestimated by nontraders, an explicit example of the consequences of a time delay may act as a corrective. Suppose the shipment of a consignment valued at $10,000 is delayed by one month and an exporter is forced to borrow this amount for one month longer than planned. If the bank charges 12 percent per year, the exporter has to make an additional interest payment of $100. If the exporter's gross trading margin is 5 percent (i.e., $500 on this consignment), we see the costs of delay in the proper perspective. Obviously, avoiding delayed shipment is an important objective. Similarly, early shipment, i.e., gaining a few days on the anticipated transport schedule, saves a nice sum of interest.[2]

The exporter also needs bank credit for *postshipment stocks*. However, banks finance these stocks in a different way: money is provided for each consignment separately, as we saw in Chapter 4. If the commercial banks have generally been willing to loan to most exporters, one reason is that these are *self-liquidating loans*.[3] After sale abroad, the credit is automatically repaid. International banks are even better equipped to provide this credit because they can monitor the payments made abroad, even before they are transferred to the exporting country.[4]

The terms of bank credit are better for postshipment stocks than for preshipment stocks. Banks charge a lower interest rate when they are in possession of the shipping documents, and provide more money for a particular consignment when an exporter can show a sales contract.[5] (Because of the risk of stock depreciation on pre-shipment stocks, banks rarely provide more than 70 percent of the exporters' finance requirements.) Clearly, it pays for an exporter to get his stocks (sitting in a port warehouse) loaded as quickly as possible.

The chances of intermediaries obtaining bank credit are slim. Only the bigger ones qualify, and they may have to pay higher interest rates than the exporters do. (Before 1940, trade credit from the exporters compensated for the lack of bank credit, as we saw in Chapter 9). The number of farmers that qualify for bank credit is negligible. As farmers tend to have little capital of their own, the burden of stockholding is high. In short, all factors combine to speed up the product flow. A farmer sells early to obtain cash. An intermediary delivers early to the exporter to repay his advances, and an exporter wants to ship early to reduce interest payments to his bank.

Physical Risks: Costs and Burdens

The stock-related physical risks can be divided into two types: those for which insurance companies offer cover and those that cannot be insured against. As to the first, there are the standard risks of accident, theft, or (complete or partial) destruction by fire. The opportunities for insurance in Africa resemble those elsewhere, the sector having been started early in the colonial period by foreign insurance companies and developed since. (Foreign companies are still predominant, although in the 1960s, they were joined by some local companies.) In principle, access to insurance is available to all actors in the channel, but in practice, all farmers and the smaller intermediaries are excluded because insurance agents are reluctant to visit small and remote customers.[6] Thus, most of the upstream stocks remain uninsured. The best protection for the owners is to sell and deliver them quickly to the larger downstream actors, who can obtain insurance coverage.

For the ocean voyage, there is marine insurance with generally lower premiums than for on-land insurance. It is therefore in the interests of an exporter to load his stocks as early as possible to come within the orbit of marine insurance.

Among the risks for which the insurance companies offer no cover, the main one is that of quality deterioration. In the humid areas of West Africa mold has always been a serious problem.[7] In the drier areas of Tropical Africa such as the Sahel, stocks of groundnuts and seed cotton have often been stored in the open air. Such stocks risk quality deterioration from even one unexpected rain shower. Other quality deterioration may be due to insects and rodents. Shrinkage (weight reduction because of loss of moisture) is also feared.[8] It is the job of intermediaries and the exporters' up-country employees to reduce all these risks by careful management. It has bred a particular style of alert management, constantly obliged to take precautions. A person has no peace of mind until the stocks are delivered to the next actor—a powerful incentive to make haste.[9]

Storage: Costs and Burdens

Since hired storage is only available in the ports, only exporters can make use of it. Although the fees are not high, they encourage exporters to keep the storage interval brief. In the remainder of the upper section, the actors have to provide their own storage. As these stores are used during only part of the year, the costs of storage cannot be accurately quantified. Since the storage burden is the same, regardless of whether the store is full or empty, it may not act as an incentive for moving stocks downstream.

The Price Risk Burden

We may be brief about the price risk because it was extensively discussed in Chapter 7 in connection with the exporters. The price risk is also faced by the intermediaries and the farmers, but the average intermediary is much less worried about it than the average exporter, mainly because the stockholding interval is much shorter. It is further relevant that during periods when the domestic prices were stabilized by EMBs, neither intermediaries nor farmers were subject to the price risk.

Since insurance companies have never provided cover for it, it is the owner who has to manage the price risk. The favorite strategy has been the downstream shifting of stocks. The intermediaries try to speed up transport so as to deliver as early as possible to the exporters. The exporters are eager to speed up transport to the port of shipment, because, simplifying the argument of Chapter 7, international selling is easiest when the product to be sold is ready for prompt shipment.

The Microview: Summary

After this review of the various categories of costs and burdens, it is clear that all actors have a strong incentive to move their stocks in a downstream direction. True, the relative weight of the categories is not the same for all actors and may vary over time, but the combined incentive is always strong. Thus, in the upper section, a high evacuation rate is always pursued. It follows that the farmers at the PMPs and the intermediaries at the IMPs are unhappy when they have to wait long before they can deliver their products. Research on whether farmers and intermediaries have indeed encountered this problem is desirable.[10]

For the exporters, there does not exist a single point that closely corresponds to either the PMP or the IMP. Normally, an exporter's title to the consignment is transferred only at the CIF point. Only then do the physical risks and financial burdens come to an end, but, as we saw, these burdens are reduced earlier, namely when the consignment passes the FOB point. The insurance companies and the banks then apply different rates which are generally lower. Another feature of CIF contracts is that the price risk comes to an end before the product is shipped. However, we recall the discussion in Chapter 7 where we saw that the sale may take place long before shipment. The economic stocks are then being moved downstream even faster than the physical stocks.[11]

What are the consequences of rapid evacuation for prices? The answer depends on the bargaining position of the actors. As the upstream actors tend to be eager to sell because of the burden of stockholding, their negotiating position is weak. The downstream actors are willing to buy, but there is no similar eagerness. As a result, their negotiating position is stronger. It is therefore plausible that there will be a discount compared with the equilibrium price. Rapid evacuation therefore tends to be accompanied by what I propose to call an *eager-seller discount*.

For a proper perspective on this discount, three comments have to be made. First, it must remain a theoretical construct because world price fluctuations make its measurement difficult, if not impossible. Second, for any welfare conclusions to be valid, it is necessary to establish that the actors are free in their decisions. Perhaps the farmers are not free but compelled by circumstances such as rural indebtedness. With the word "eager," the voluntary character of the decisions is stressed, not only of the farmers but also of the intermediaries and exporters. (Needless to say, compulsion would further accelerate the flow.) Third, there has always been a school of thought that recommends less eagerness to the farmers and thus later sales. However, this recommendation is unwise if it increases the stockholding burden to such an extent that it outweighs the higher revenue due to a later sale.

Major Infrastructural Improvements

Until now, we studied produce evacuation as a human effort within the existing infrastructure. But what happened in periods when the infrastructure was drastically improved? In those periods, the goal of rapid evacuation operated even more powerfully, as may be illustrated with the following brief accounts from Nigeria, Senegal, and Uganda. In the 1910s and 1920s, the evacuation programs in those countries switched from being based on cheap, but slow river (or lake) transport to fast, but expensive rail transport. In Nigeria, rail and river began to compete in 1912 when the railway line from Lagos to Kano in the heart of Northern Nigeria was opened. Until 1912, all export crops from this area had been carried on the Niger River to ports of shipment in the Niger Delta, but with the railway came a diversion of exports from the Delta ports to

Lagos. Several exporters responded by moving their headquarters from the Delta to Lagos.[12]

In French West Africa, competition between rail and river began in 1923 when the railway line reached Kayes, an inland port on the upper Senegal River. Export crops from the French Sudan (now Mali) had until then been shipped from Kayes. Unfortunately for the exporters, each year, stocks had to lie there in storage for many months, as evacuation by river was restricted to a few weeks when the Senegal was in full spate. This delay disappeared after 1923 because exporters could rail their crops to Dakar soon after they were purchased.[13] In Uganda, there was a comparable decline of water transport on Lake Victoria (Ford, 1955) after the railway line from Mombasa was extended to Kampala in 1928 (Hill, 1949, p. 473). The fact that several exporters in these countries were willing to pay more for the faster form of transport proves to me that they pursued rapid evacuation.[14]

THE MACROECONOMIC VIEW

We now turn to the macroconsequences of rapid evacuation. Once again we begin with the stocks and the stockholding burden, but instead of looking at the stocks of the individual actors, we consider the aggregate stocks in the upper section. Since the stockholding burden depends on the stocks, it is important to have a rough idea of the way the sector's stocks increase and decrease each year. In the simple model used here harvesting is responsible for stock growth and shipments for stock decline. This leads to the triangle AGF shown in Figure 10.1—an approximation of the real curve which tends to have a hump shape. It is further assumed that

FIGURE 10.1. National Stock Cycle of Commodity X

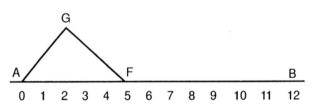

(1) the harvest season lasts two months; (2) the buying season coincides with the harvest season; (3) there is no local consumption; and (4) shipments begin during the buying season and continue until the whole crop has been exported. In the diagram, the aggregate stocks peak at the end of the second month[15] and fall to zero at the end of the fifth month. There is an off-season (the period between points F and B) of seven months in which no stocks are available in the country.

Since all actors try to reduce their own stocks, they contribute to the shrinking of the triangle: it becomes lower and narrower. The combined efforts cause point F to shift to the left. Whether the shift is significant depends on several factors, most of which have already been discussed. However, some factors, typically of a national nature, have to be briefly mentioned here: the collaboration of the shipping companies (Do they adjust their shipping space to the needs of the sector, for instance, when there is a bumper crop?); the capacity of the port and the occurrence of port congestion; and the threat of dockers' strikes.

What are the macroadvantages of shifting point F to the left? The answer directly follows from the earlier discussion. The stockholding burden is reduced for the whole upper section and indirectly for the country. In a country where capital is scarce, the insurance sector underdeveloped, and the climate inimical to storage—conditions applying to many African countries—it makes sense, from a national perspective, to shift the stockholding burden to actors abroad. However, there are two disadvantages: (1) there is a national equivalent of the eager-seller discount and (2) the country does not benefit from any stock appreciation which happens to occur during the period between points F and B. It is significant that over the years, those actively involved in the sector have been consistently in favor of an early point F. Apparently, they felt that the advantages of rapid evacuation outweighed the disadvantages. Economists should give due attention to this view.

If the goal of rapid evacuation was well known in West Africa, was it perhaps weaker or absent elsewhere in Africa? Was it, for instance, weak in countries with many grower-exporters? As the cumulative aspect was missing there, we must envisage the possibility that grower-exporters have been less interested in early evacuation than trader-exporters. It is, for instance, plausible that grower-exporters with privileged access to capital and bank credit in the countries of destination, such as many foreign plantations enjoyed, could afford to wait.[16]

Registering the Seasons

In the commodity sectors discussed here, it is relatively easy to identify five seasons: the *harvest season*, the *buying season*, the *processing season*, the *internal evacuation season*, and the *shipping season*. In many African countries, these concepts are familiar to people. They are not only discussed in specialized journals, but also in the general newspapers. The harvest season is determined by the climate, which ensures great similarity over the years. The buying season[17] coincides with the harvest season unless there is on-farm processing. (We assume that farmers sell as early as possible.) The processing season is only relevant for crops such as cotton, which require off-farm processing. The internal evacuation season begins at the same time as the buying season, but may continue much longer, particularly in countries where (a) the transport capacity is inadequate; (b) the production areas are remote; and (c) the evacuation season coincides with the rainy season. The shipping season may be defined as the interval between the first and the last shipment of the current crop. It may begin early, but in many countries there is a time lag between the beginning of buying and that of shipping. The time lag may be caused by processing or geographical factors (e.g., some production areas such as the groundnut areas in the Sahel are remote). The time lag tends to shift point F to the right.

In view of the economic significance of rapid evacuation, I plead for the systematic registration of the seasons just mentioned. Registration helps to measure a sector's physical performance. True, point F will not fall on the same date each year, but it is useful to examine the factors responsible for deviations. In this way, any slackness of one of the many organizations involved can be identified early. The central banks seem to me the most suitable organization for assuming responsibility for this registration. This is so because in many countries, the banks have indirect information about the sector's stocks—indirect because some people in the banking system are in a position to aggregate the data on seasonal credit requests.[18] Moreover, the central banks register currency in circulation, a variable which in many countries corresponds to the stocks of the AGF curve.[19]

I further plead for the registration of a sixth season, the *external payments season*. This suggestion is based on the following considerations. Each consignment that is shipped is followed by a payment, which is not due until the product has arrived in the port of destination. As these payments, like the shipping dates of the consignments, are spread over a few months, the external payments season must be defined as the interval between the first and the last payment. (A complication is caused by the variation in the length of the sea voyage: importers in remote ports of destination pay late.) This suggestion does not involve much additional work because in many countries, a simple system for monitoring the external payments is already in place.

The Macroeconomic View: Global

What are the external repercussions of rapid evacuation? Is it possible to trace them in the lower sections of the TOMCs, that is, in the countries of destination? And, if all the countries in which the commodity is grown are less-developed countries and characterized by rapid evacuation, is there a truly global effect? Below, I develop a simple global model to address these questions.

The new model is illustrated by Figure 10.2, which shows the global stocks of commodity X. The following assumptions have been made. First, there are a dozen exporting countries, all of them LDCs. Second, they all have the same harvest season.[20] Third, evacuation takes place in such a way that the shipping seasons for

FIGURE 10.2. Global Stock Cycle of Commodity X

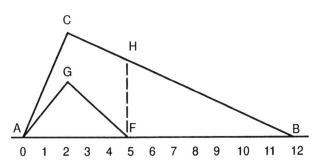

all countries coincide. As a result, the triangle AGF has the same shape as in Figure 10.1, but of course, the volume is many times higher. Fourth, global demand equals global supply: there are no carryover stocks and no stockouts at point B. Fifth, global demand (from consumers and/or manufacturers) is constant over the year: the slope of the line CB is constant. Sixth, there is no demand in the countries of origin (which was realistic in Africa until about 1960; for subsequent changes, see below). Seventh, stocks which are afloat (i.e., in trans-oceanic transit) are zero: all stocks are supposed to be on land, either in the countries of origin or in those of destination. As with all agricultural products, supply cannot easily be adjusted to demand, leading to large seasonal stocks (Tousley, Clark, and Clark, 1962, p. 478).

In the model, there is a pronounced accumulation of stocks in the countries of destination, represented by the area bordered by the lines AC, CB, BF, FG, and GA. To the right of the line FH, all stocks are at destination. To the left, they are partly at origin and partly at destination. It is important to realize that these stocks are not yet needed by the manufacturers. Thus, *global wholesalers* are needed, entrepreneurs who are willing to hold the stocks which the exporters no longer want and which the manufacturers do not yet need.[21] The contrast is interesting. For the exporters (and many manufacturers), stockholding tends to be an unwelcome but unavoidable burden, but for the global wholesalers, it represents a function in which they deliberately specialize.

A supplementary conclusion is that it is misleading to speak of the average velocity of the product flow for the entire TOMC. At most, the average velocity can be used as a basis for comparison, as is done in the following observation. The velocity (averaged for the section) is high in the upper section; it is standard (and beyond control of the stock owners) in the middle section;[22] and it is slow in (parts of) the lower section. Unfortunately, foreign trade statistics are unsuitable for measuring these differences in velocity.

The Merchant Complex

It is helpful to explore the relationship between this model and the literature on commodity merchants on the supposition that the model's global wholesalers correspond to the international commodity merchants. The latter are known to even out imbalances between global production and consumption, not only the seasonal ones (on which the model focuses), but also those caused by booms and recessions, rapid changes in consumer preferences, and improvements in processing technology.

Most of the merchant stocks are held in the major ports of destination. In these ports, as we saw in Chapter 3, there are large, specialized warehouses for coffee, cocoa, cotton, etc. The storage costs and the physical risks are taken over by the warehouse owner, who in turn charges a standard rate per ton and per day. The storage costs of stocks sitting in these warehouses are very low.

Stocks at these warehouses are an important item in the global statistics for various commodities. However, when interpreting these statistics, we should be aware of two pitfalls. They exclude the stocks afloat on the oceans[23] and do not indicate whether economic stocks deviate from the physical ones. In Chapter 7, we saw that the exporters in Africa generally sell before shipment, and occasionally much earlier. Insofar as the merchants directly buy from these exporters, their economic stocks greatly exceed their physical stocks. Put more generally, the actors in the lower section own nearly all the stocks afloat and part of the physical stocks in Africa. We now see that a major subject of Chapter 7, the downward shifting of the price risk, is one element of the subject of the present chapter, the rapid shifting of commodity stocks within the TOMC.

The merchants are prepared to assume the stockholding burden because of the profits they expect to earn. Broadly speaking, they may earn two types of profit. The first is a simple wholesaler's profit, justified by the services of evening out peaks of supply and demand. The second type of profit concerns the windfall profits resulting from stock appreciation during periods when commodity prices rise rapidly. The merchants are often criticized on this score, but it is doubtful whether they themselves make these profits. Since they cover their long positions by hedging operations on the terminal markets, most of the windfalls go to what are called speculators. These are criticized even more than the merchants, but in fairness, it should be said that the speculators as a group at times incur large losses—comparable in value to the windfalls.

Apparently, the counterpart of rapid evacuation is not a single category of actors, but a whole *merchant complex*: the merchants themselves, the operators of the technically advanced warehouses, the banks who provide credit to finance the merchant stocks, the terminal markets which enable the merchants to hedge their risks, the speculators which make the terminal markets work, and the analysts who systematize market intelligence. No category of actors can be missed.

The extensive theoretical literature on stockholding is largely based on the operations and strategies of the merchants. They use sophisticated calculations in which storage costs, changes in interest rates, the costs of hedging, and expectations about prices and exchange rates play a role. Theories on the time patterns of the commodity prices are a basic element of the literature (Williams and Wright, 1991). However, the literature on stockholding does not seem to apply to the exporters in Africa. It would appear that these exporters deliberately close their eyes to opportunities to profit from stockholding and try to get rid of their stocks as early as possible.

The question of why the merchant function is concentrated in a number of developed countries can also be explored from a macroeconomic perspective. One explanation is that the geographically uneven distribution of the global stocks reflects differences in factor endowment, that is, the relative scarcity of capital in the LDCs compared to the capital abundance in the (developed) countries of destination. An additional explanation must be sought, I feel, in the differing capacity and willingness to assume the price risk burden. This combination is found in only a few developed countries.[24]

RAPID EVACUATION: AN APPRAISAL

Do we have to revise our earlier opinion about rapid evacuation now that we have identified the merchant complex as the counterpart of rapid evacuation? We have now concluded that the actors in this complex benefit from the eager-seller discount and from price rises during the period they own the stocks.[25] In return, the actors have to bear more risks and to provide more finance.

It is clear that rapid evacuation and the accompanying concentration of global stocks is the result of the free play of economic forces. For classical economists, this is sufficient reason to assume that this global phenomenon is an efficient and beneficial solution for all actors in the TOMC, including those in Africa. I say "assume" because objective measurement of the advantages and disadvantages seems impossible. Classical economists should not only approve of rapid evacuation, but should also warn against measures that obstruct it. However, there are two conditions. First, as noted before, the farmers must be truly free when selling and governments should carefully monitor and guard farmers' freedom. Second, market intelligence must be public because secret knowledge can be used by strong actors at the expense of weaker actors. There is general consensus that this condition is fulfilled in the case of the commodity exchanges.

Unfortunately, the process of rapid evacuation cannot easily be incorporated into the theories of international trade which are based on production costs. How do we incorporate the stockholding burden? Should it be included in the theoretical world price? These questions have to be faced by all theories which study the international terms of trade. The Dependency and the Center-Periphery theories, for instance, argue that there is a systematic depressing effect on the prices of agricultural products grown in the Periphery due to the "inferiority" of agriculture in relation to industry. I also see a depressing effect on prices, but I attribute it to the eager-seller discount. In this connection, I must point out that the Center in my analysis is not characterized by its industrial predominance, but by its commercial sophistication—a sophistication which extends to the ports of shipment in the Periphery!

How one views rapid evacuation will color a person's opinions about the commodity merchants. Many African governments have tended to be critical of them, which was a major factor in some of their policies (see below). In contrast, the governments of the countries where the complex was located, such as the United Kingdom and the United States, have tended to approve of merchants and tried to foster the complex by adopting liberal policies with regard to imports and foreign exchange, by approving bonded warehouses, and so on.

Although the national eager-seller discount is no more than a theoretical construct, it may help in policy discussions. We may, for instance, ask whether farmers, traders, and governments think about it in the same way. It seems to me that farmers and traders object much less to the discount (i.e., that part of the discount which they themselves lose) than governments. This probably reflects a different time perspective, with governments having the longer horizon. One consequence is that an African government that wants to counteract rapid evacuation may not find support, but obstruction among the farmers.

The Burden Blindness of the EMBs

We now return from the global to the African perspective. Some years ago, it struck me that so little had been written about rapid evacuation since World War II. Indeed, it was seriously underrated for several decades. In fact, I go further by postulating that nearly all the post-1945 EMBs were more or less blind to the stockholding burden.[26] Because of their public status, sanctions were practically absent. If large sums of money were lost because of this blindness, bankruptcy was nearly always averted by the government.

The commercial banks (and in some countries, the central bank) did not help to correct the vision of the EMB managers. The banks were far too generous with credit. The self-liquidating characteristic of seasonal credit and the EMBs' public status made the EMBs ideal clients. Insofar as the banks failed to insist on alert cash flow management, the EMBs lacked the discipline of a critical lender.[27] It further seems that consultants had no correcting effect. Consultants tended to focus on prices and marketing costs. To my mind, the physical performance of the EMBs should also have been evaluated and measured.[28] Were exceptionally long trading seasons prevented?[29] Research on EMBs ought to be double-edged: measures to raise selling prices as well as measures to lower the stockholding burden should be considered. Success in either direction was favorable for the farmers.

By contrast, the private licensed buyers appointed by the EMBs were well aware of the stock-holding burden and pursued the goal of rapid evacuation. Often they were the only ones who sought rapid evacuation, as the EMBs and the cooperatives were blind or slack in their management. I expect that there were two effects on a sector's seasonal finance requirements. First, the amount increased because stocks were shipped late and, second, the private licensed buyers tried to decrease their own share of the total amount, usually at the expense of the national EMBs.[30]

In the francophone countries where private exporters were retained, an awareness of the stockholding burden never entirely disappeared. Private exporters complained bitterly when bureaucratic procedures caused delays in shipment. It is further interesting to note that two French companies have established warehouses in several ports of francophone Africa and issue warehouse receipts for the commodity stocks deposited by the exporters. These stocks awaiting shipment do not lie long in these depots, but even so, the receipts reduce the financial problems of the exporters by enabling them to obtain bank credit on better terms.[31]

Export Quotas

The imposition of export quotas formed an integral part of agreements such as the International Coffee Agreement. Under this agreement (established in 1962), export quotas were imposed whenever world coffee prices were low. Market theory suggests that a temporal reduction of global supply by means of export quotas raises prices in favor of the producing countries. Support for this strategy has been widespread, both in Africa and abroad. However, the analysis in this chapter shows that there is a drawback to this strategy. When an export quota operates, part of a sector's stocks is retained in the country of origin and the stockholding burden has to be borne much longer than in a free market.[32] (Remarkably little has been written about this consequence; one reason may be that it was the EMBs which held the excess stocks and that their managers were "burden blind.") Here, we see that market theory recommends export quotas, while MCA expresses doubts about them. Further analysis is necessary to resolve this conflict. It must further be pointed out that the drawback of export quotas is negligible when they are accompanied by production quotas. In those instances, no stocks have to be held back.

In 1989, after the collapse of the International Coffee Agreement and the suspension of the export quotas, the goal of rapid evacuation reemerged powerfully. Within a short time, the coffee producing countries were shipping the coffee stocks that had been dammed up by the quotas in addition to shipments from their current harvests. The result was an unprecedented fall in world prices. These low prices brought a reaction in favor of new export quotas, which, indeed, were adopted in 1993 under the Coffee Retention Scheme, an agreement among producing countries only. For our purposes, the question of whether the consuming countries are party to the agreement or not, is immaterial. Those setting an export quota should, in either case, be vividly aware of the burdens involved.

Vertical Diversification

Since about 1960 many macroeconomists have recommended to African governments that they promote the processing of products that were formerly exported in the raw state. Examples relevant to our argument are the grinding of cocoa, the crushing of oil seeds, and the roasting of coffee. These activities led to exports of cocoa mass, vegetable oil, and roast coffee respectively (Dijkstra and Van der Laan, 1990). There is general agreement about the macroeconomic advantages of *vertical export diversification* (as this type of industrialization came to be called): the country of production obtains a higher price and more people are employed.[33] In the chorus endorsing industrialization of this type, I strike a different note. There is a serious and often overlooked disadvantage: the stocks in the channel are held for a longer period in the country of origin because the processed product is now shipped later than the unprocessed product was previously. Total stocks awaiting shipment (consisting of both processed and unprocessed product) are therefore larger than before. One could say that part of the global merchant stocks are now held in the countries of origin. Accordingly, the burden of stockholding at origin increases. Once again, market theory and MCA come to conflicting recommendations.[34]

Glossary

burden: Here defined as a stock-related cost item that cannot be measured, either directly or by means of opportunity costs.

buying season [of crop X in country Y]: The interval between the day on which buying at the PMPs begins and the day on which it ends.

(N.B. The official buying season of the export marketing boards tends to be longer.)

eager-seller discount: A discount which arises when the seller is eager to sell quickly, while the buyer is not eager to buy.

external payments season [of crop X in country Y]: The interval between the day on which the first external payment is received and the day on which the last one is received for a certain year's crop.

first stage of evacuation: The evacuation of the harvest from the PMPs to the port of shipment.

global [commodity] **wholesalers:** Entrepreneurs who hold stocks of a commodity. These stocks come into being because the exporters no longer want them and the manufacturers do not yet need them.

harvest season [of crop X in country Y]: The interval between the first and the last day of harvesting crop X.

intermezzo storage: Short-term storage at the PMPs and IMPs, usually in connection with the transfer of the produce from one mode of transport to another.

internal evacuation season [of crop X in country Y]: Period during which internal evacuation, from the PMPs to the port of shipment, takes place.

merchant complex: An economic complex based on the cooperation of merchants, public warehouse owners, commercial banks, terminal markets, speculators, and market analysts.

postshipment stocks: The part of the exporter's stocks that has already been shipped.

preshipment stocks: The part of the exporter's stocks that has not yet been shipped.

processing season [of crop X in country Y]: The interval between the first and last day of the industrial processing of a crop during a given year.

rapid produce evacuation: The rapid removal of the current harvest of a particular export crop from its country of production.

seasonal finance requirements: Here defined as the amount of short-term finance which a trader (exporter or intermediary) needs during the buying season.

second stage of evacuation: Period during which the produce in the port warehouses is prepared for shipment and shipped.

self-liquidating loans: Short-term loans in connection with stocks of highly marketable products. These types of loans are attractive to the lender because the borrower will soon turn his/her stocks into cash.

shipping season [of crop X in country Y]: The interval between the day on which the first consignment of a certain year's crop is shipped and the day on which the last consignment leaves the country.

stock owner: Here defined as the actor who owns stocks of trans-oceanic agricultural commodities. Normally the stock owner is an organization, often a large one because the trade in these commodities is complex.

total burden: Here defined as the total costs of stockholding, including non-measurable costs.

vertical export diversification: A change in a country's export basket resulting from the substitution of processed products for raw materials; it requires a special form of industrialization which converts raw materials into processed products.

Chapter 11

Minor Crops and Their TOMCs

After having studied the commodities in the last four chapters, we now turn to the entire-channel crops (Chapters 11 through 13). Exporters who sell these crops have to study the entire marketing channel, including the lower section. (As we saw, the commodity exporter could ignore this section.) The better exporters know the actors and intricacies of the lower section, the higher their price will be. Put differently, the exporters of an entire-channel crop have to make a determined effort to get their product sold: exporters have to invest in marketing skills and promotional activities. The term "marketing" in these next three chapters thus has its customary meaning of marketing management.

THE LOWER SECTION

The main actors in the lower sections of the TOMCs for minor crops are manufacturers and merchants. Some exporters in Africa are able to sell directly to a manufacturer, but the others have to sell to merchants. These importer-merchants stand between the exporters in Africa and the manufacturers. The importer-merchants hold stocks, but their operations are more limited than those of the commodity merchants: they cannot hedge their price risks because no terminal markets exist for these crops. The importer-merchant's reluctance to accumulate large stocks is therefore greater. The importer-merchants are nevertheless important because they help provide a somewhat greater degree of certainty, continuity, and marketability. This is particularly valuable in consolidating a channel where there are many small actors in the lower section.

The TOMCs with importer-merchants nearly always form a conventional marketing channel. Insofar as vertical coordination exists, it is associated with direct transactions between exporters and manufacturers. The best chances for vertical coordination occur when an exporter is a grower-exporter (see below).

When studying the minor crops, we must be prepared for more variation in the type of collaboration between exporters and importers than found in the commodity TOMCs. We must further be aware of the power of circumstances in promoting (or disrupting) such collaboration. Various circumstances which may favor the creation of a new TOMC are: temporary prosperity and new culinary fashions in the countries of destination, new technologies, and the threat of war. However, the TOMCs remain fragile, which is confirmed by studies in which the lack of channel continuity is deplored. Since the channels are more easily disrupted than commodity channels, many exporters must spend a great deal of time in maintaining and consolidating their TOMCs.

The absence of a physical world market, the weakness of trade associations, the paucity of public price information, and the poor or nonexisting grading systems create a situation which tends to increase the role of brokers. The brokers in the major ports of destination are the principal sources of inside information for the exporters.

Below, we will discuss a dozen crops, divided into three categories: (a) those produced by large growers, (b) those grown by small farmers, and (c) miscellaneous crops. The approach is analytical and general, but a fair amount of detail is included. The historical details provide a fuller understanding of the TOMCs—in several cases, they have early origins but checkered fortunes—and help to point out the sources of their vulnerability to disruption.

CROPS PRODUCED BY LARGE GROWERS

Sisal is typically a plantation crop because a large installation is necessary to extract the fibers from the leaves. The first plantation in Africa was established in German East Africa (later Tanganyika) in the 1890s. In the following decades, sisal cultivation spread to Kenya and Mozambique. The sisal TOMCs have always been simple because one category of actors, the rope manufacturers, have dominated the lower section.

In the interwar period, some sisal plantations in East Africa sold directly to the manufacturers by means of forward sales. The other sisal producers sold to merchants operating in the world market for hard fibers in London.[1] They entrusted the actual shipping and selling to commission agents with offices both in Africa and in London (Stahl, 1951). Direct sales to a manufacturer offered better results (good prices and early sales), but it seems that this option was only open to plantation companies who produced good, consistent quality. A relationship of trust developed between them and the rope manufacturers, but farther-reaching forms of channel coordination did not seem necessary. During World War II, the marketing and shipping of sisal from Tanganyika and Kenya was centralized (Westcott, 1984).[2]

Two other products—pyrethrum and wattle bark—have also been typically East African products. They were not grown on corporate plantations, but by large farmers and estate owners. Like the corporate plantation managers, these growers had the commercial sophistication necessary to understand the problems of international marketing. Even if they themselves did not organize the export activities, they were able to assess the performance of those who did. In the colonial period all of the large growers were non-Africans, with the settler farmers in Kenya and Tanganyika being the most conspicuous group.

The flowers of pyrethrum, a variety of chrysanthemum, contain a natural insecticide, which is harmless for human beings. Around 1930 some settler farmers in Kenya began to grow pyrethrum for export. This involved the picking (which is labor intensive), drying, and baling of the flower heads. Since 1945, when an extraction factory was established, Kenya's exports have consisted of both flower heads and extract. The two factories (near Nakuru) are run by the parastatal Pyrethrum Board, in whose operations the trading company of Mitchell Cotts has a minority interest (Heyer, 1976, p. 347). The pyrethrum sector continues to be important in Kenya and Tanzania, but African growers have largely replaced the settler farmers.

The wattle tree is grown for its wood (for local use) and for the bark which is rich in tanning materials. At first the bark itself was exported. Exports from Kenya began in 1910. In the 1930s processing facilities were installed in that country in order to produce and export wattle extract rather than bark. This was favorable for the settler farmers who considerably expanded their production before World War II. After the war the share of the African growers in total production increased, a process which was accompanied by a decrease of the average size of the plantations. By 1970 there were two extraction factories in Kenya and one in Tanzania (Acland, 1971, p. 231). Each factory buys from many growers. The wattle sector has always been left to private enterprise.

CROPS GROWN BY SMALL FARMERS

Many minor crops have been grown by small producers. Examples are ginger from Nigeria and Sierra Leone, castor seed from Tanzania and Uganda, vanilla and pepper from Madagascar, and burley tobacco from several East African countries. (The cheap burley tobacco is to be distinguished from the more expensive Virginia tobacco discussed in the next chapter.) We first consider the lower sections of the TOMCs. Castor seed is industrially crushed by manufacturers to provide an engine lubricant; piassava, a hard fiber, is bought by brush makers, and so on. As ginger is used only in small quantities in bakery and confectionery products, demand is fragmented and creates a niche for merchants. The TOMC for ginger is a long one, with at least seven successive actors: small farmer, intermediary, exporter, merchant-importer, wholesaler, baker, and consumer.

In the upper sections, there have been trader-exporters who bought these minor crops from the farmers (often via intermediaries) and took care of the exportation of these crops. Something more may be said about the trader-exporters in West Africa. The European trading companies (studied as commodity exporters in Chapter 9) also handled many minor crops, but they treated them as sidelines. Trade in minor crops was not a goal in itself, but a useful way of consolidating rural contacts and of raising local purchasing power (which enhanced the companies' sales of merchandise).

Occasionally, exporting these crops was a successful activity for smaller businessmen. During my research in Sierra Leone, I was told that a French trader had made his fortune in the ginger trade in the 1920s, not in competition with the big companies, but simply because the latter were not sufficiently interested. In the same way, a few African and Lebanese traders became exporters in the piassava sector in Sierra Leone around 1930.[3] These traders were quick off the mark and southern Sierra Leone experienced a "piassava boom" in the early 1930s, at a time when most of West Africa was suffering from the global economic depression (Van der Laan, 1975, pp. 49–50).

It was natural for the EMBs of the postwar period to take over the trade in the minor crops, but it is my impression that they were less active in this area than the European trading companies before them. Whereas the latter had used these crops to strengthen their general position and to raise rural purchasing power, the EMBs, not threatened by competitors, could ignore these considerations. However, even more important, the EMBs were designed to sell commodities and were therefore not well equipped to sell minor crops. It seems that many EMBs were reluctant to expand their marketing departments for this purpose.[4] In Sierra Leone, a noticeable reluctance on the part of the national EMB was observed with regard to minor crops such as ginger and piassava. The politicians sometimes forced the EMB to assume control of them, but after some time this policy would be relaxed and the export of these crops would be handed back to the private sector (ibid., p. 65). In Ghana, the cocoa EMB was instructed to handle some ten additional export crops in 1961, but the effect on the board's actual activity was negligible (Arhin, 1985, p. 41).

Another postwar market development was that certain oilseeds began to be valued in their own right and not only as a raw material for the extraction of oil. Thus, the larger and more regularly shaped groundnuts (known by insiders as HPS, for Hand Picked and Selected) were demanded by confectioners, and sesame seed was bought for its special flavor. This differentiation of demand removed part of the crop from the bulk commodity trade, and turned it into a minor crop. The importers' special requirements now had to be communicated to the exporters and through them to the small growers. This proved difficult, even if the exporter was prepared to pay a considerable premium for the desired quality, and even if officials of the department of agriculture actively cooperated, as in Uganda (Martin, 1963, p. 9).

DIFFERENCES WITH THE COMMODITIES

Although the minor crops are less marketable than the commodities and their world markets are less well organized, there are two common features: the TOMCs of both types of crop are Conventional Marketing Channels and importer-merchants are important actors. The question then arises of whether the two types of crop differ in kind or in degree.

Most economists are inclined to speak of a difference in degree and thus recommend the same marketing institutions for both types of crop, that is, economists recommend that commodity exchanges be established for as many of the minor crops as possible.[5] While several efforts in this direction have been made, nearly all have failed. The general explanation has been that the number of transactions in the new exchange turned out to be too small for success.

I contend that the two types of crop differ in kind. When the number of (actual or potential) international traders for a particular crop remains below a certain level, the conditions for the establishment of a commodity exchange are not fulfilled. In this situation, the concepts developed by Williamson (1975) are useful. He argues that "small numbers" in a market facilitate "opportunistic behavior," while large numbers put an effective curb on such behavior. The world markets for the minor crops appear to be characterized by such small numbers.

The following situation may serve as an example of opportunism or *disloyal behavior*—a term which may be more familiar among traders in Africa: a particular exporter and an importer-manufacturer have been trading for many years and consider each other long-standing trading partners. However, as the relationship is not based on a long-term contract, loyalty cannot be enforced in any way. Disloyalty occurs when the importer-manufacturer secretly scouts around for a cheaper supplier, possibly in an entirely different country. When the importer-manufacturer ceases to buy from his regular supplier, the latter is greatly (and often suddenly) disappointed in his expectations and has to search for another buyer-importer. The costs of buyer search, one of the elements of transaction costs, are unpredictable. If the exporter is lucky, he may incur few marketing costs before locating a satisfactory buyer. However, he may also be unlucky. Here is a significant difference from a situation with a commodity exchange where an exporter hardly suffers when a long-standing buyer ceases to buy from him. As an exporter can easily find another buyer, switching from one supplier to another is not frowned upon in the commodity exchanges. There is no virtue in, or premium on, loyalty.

A crucial point in Williamson's argument is that given a situation of small numbers and opportunistic behavior, businesspeople tend to opt for vertical integration as the best offsetting solution. In the TOMCs for minor crops, a similar tendency toward vertical integration and coordination across the ocean can be discerned. Exporters see it as a valuable safeguard against trans-oceanic disloyalty. My (African-based) impression is that coordination across the ocean has hardly ever been realistic for the minor crops.

Losing one's long-term buyer-importer can happen at any time: it is not a special feature of depression years, but, of course, in such a period, the repercussions seem more serious. In the same way, the repercussions may have been largely disregarded in times of prosperity, such as in the first two decades after World War II.

The problems that occur when a buyer-importer turns to another source of supply are, unfortunately, not limited to the exporter. If, for instance, a trader-exporter is unable to find another buyer-importer and stops buying this crop from the farmers, the farmers may be forced to abandon the crop, and the sector collapses. This throws a new light on the vulnerability of the minor-crop sectors.

MISCELLANEOUS CROPS

Since the minor crops are a residual category, it is not surprising that the four products which conclude this chapter have little in common: cloves, cashew nuts, pre-1925 cocoa, and canned fruit.

Cloves and cashew nuts can be grouped together because they have both figured in an Asian TOMC, linking producers in East Africa with importers in South and Southeast Asia. Long before the colonial period, there were clove plantations on Zanzibar and

Pemba, operated by Arab grower-exporters. To European observers, these businessmen seemed small, but they deserve respect because they had succeeded in establishing a TOMC merely on the basis of sailing ships.[6] This TOMC continued under colonial rule with the steamers of European shipping lines replacing the sailing vessels. The Asian exporters operated without serious competition from European businessmen, who were not interested in this product unless it was shipped to Europe.[7]

Cashew nuts have long been grown in Tanganyika and Mozambique, but in contrast to cloves they are produced on a small scale. In the interwar period Indian entrepreneurs began to export the raw nuts to India where they were decorticated. The kernels were then reexported, mainly to the USA—a double TOMC. After World War II, Tanganyika/Tanzania and Mozambique experienced a cashew nut boom which continued till the mid-1970s. In 1954, a private decorticating factory was established in Mtwara, Tanganyika (Hailey, 1957, p. 842). In 1962, the Tanzanian government established an EMB for the cashew nut sector. After some years this EMB decided to set up its own decorticating factories and to export the kernels so obtained directly (a single TOMC) to the countries of consumption. Twelve factories were built, but as their performance was poor, the export of raw nuts to India continued, albeit on a reduced scale. After the liberalization of the cashew nut trade in 1991/92 the question arose whether the new private exporters would be interested in buying or leasing some of these factories (Jaffee, 1995). The answer is still uncertain.

Cocoa was discussed as a commodity in the previous chapters, but it should be remembered that cocoa was a minor crop until 1925, when the first commodity exchange for this product was opened in New York. Researchers studying the early years may therefore find the analysis in this chapter helpful. Since it was then more difficult to establish contact between the growers in Africa and the chocolate manufacturers in Europe, we should pay special attention to actors that took initiatives in that period. Well known is the decision made by Cadbury, the chocolate manufacturer in Birmingham, to establish a buying organization in West Africa in 1906 (Southall, 1978). The Cadbury employees could talk to the farmers, indicating what quality the company wanted, and could reward the production of superior cocoa with a higher price. However, initiatives were also taken by some of the early cocoa farmers in West Africa such as those in the Gold Coast and in the hinterland of Lagos (Hopkins, 1978). Many of them were sophisticated growers, not unlike the European farmers on the other side of the continent. Their attention to marketing aspects manifested itself in the formation of cooperatives, which first made consignment sales to England and later were important as suppliers of cocoa to Cadbury (Southall, 1978).

Canned fruit began to be exported from Tropical Africa in the 1950s. At first sight, it seems this product should be discussed in the chapter on perishable crops. However, our criterion for classification is whether the crop is perishable at the export stage, and this is not the case. The crucial actor is the *canner-exporter* who converts a perishable input into a durable output. Canning was one solution to the problem of the high cost of transporting perishable produce by refrigerated ships. Provided that the taste of the fresh product can be preserved, consumers may find the canned product an acceptable substitute for the fresh one. Pineapple proved to be a good candidate. In Kenya, the cultivation of pineapple, geared to a canning factory, was started around 1950 (Swainson, 1980). Del Monte came to Kenya in 1965 to manage the factory and purchased it three years later. In Côte d'Ivoire a French company started a canning factory for pineapple in 1949. The country's pineapple export has long consisted of both canned and fresh fruit. (For the latter, see Chapter 13.)

It is useful to explore the differences between a TOMC with, and without a canning factory. Downstream of the factory, there is no need to speed up transport because canned fruit does not deteriorate. There is further no risk of products spoiling when stocks accumulate. Exporters no longer stand with their backs against the wall when they negotiate with importers. Moreover, canned fruit can be carried over long distances. One could say that the international market for canned pineapple is a genuine global market, whereas the market for fresh pineapple is a regional or *subglobal market*. In some subglobal markets, competition is less strong than in a global market, and prices obtained by exporters may be somewhat higher. These points show that a TOMC for canned pineapple significantly differs from one for fresh fruit.

Upstream of the canning factory, operations are as urgent as in a TOMC without a factory. Between the moment of harvesting and the moment of boiling or pasteurization that precedes canning, every minute counts. Alert management is required in both cases.

The managers of some canning factories are convinced that operations and timing can only be successful when the fruits (or vegetables) are grown on a factory's own land, that is, on a plantation. (This was Del Monte's view in Kenya.) Others are prepared to leave the cultivation in the hands of independent farmers from whom the factory buys. In the last case, detailed agreements or contracts between the factory and the farmers are necessary. Occasionally, the two systems are combined. The factory has a nucleus plantation, good for perhaps one-third of its canning capacity. This nucleus guarantees that the factory will never be completely idle. For the other two-thirds, the factory relies on outgrowers. This arrangement has been used in Côte d'Ivoire for many years (Sawadogo, 1977).

Glossary

canner-exporter: An entrepreneur who is both a canner (of fruits and vegetables) and an exporter.

disloyal behavior: Behavior that runs counter to the expectations of other channel members. For example, importers are considered disloyal when they suddenly cease to buy from exporters who have long been their regular suppliers.

subglobal market: A separate compartment of the world market. High transportation costs, such as those resulting from refrigerated ocean transport, are one of the factors that create such compartments.

Chapter 12

Auction Crops and Their TOMCs

In Chapter 5, auction crops were defined as crops for which product differentiation on the basis of sensory (or other nonmeasurable) characteristics is profitable in the eyes of influential actors in the TOMC. It was noted that differentiation is not only profitable, but also difficult and costly. Three specific crops were mentioned: tea, tobacco, and arabica coffee.

In this chapter, the focus is on local auctions, that is, auction markets organized in Africa. In the 1930s, there were high expectations for local auctions in East Africa. In less than a decade, seven regular auction markets were set up. Unfortunately, the literature about them is limited and mainly descriptive. By the time a proper evaluation was due, World War II was raging and such evaluation fell by the wayside. After the war, the auction solution was continued as a matter of course and is still in favor today. A deep-delving economic analysis was not made until recently (Houtkamp and Van der Laan, 1993). The significance of the local auctions for modernizing the TOMCs has been largely overlooked in academic research, and the actors in Africa have been given too little credit for their initiatives.

FACTS AND VIEWS OF THE 1930s

Table 12.1 summarizes information on the early local auctions of the 1930s. A cursory glance at the table reveals that the five countries in which the auctions were established were all located on the Eastern side of the continent and were under British rule. Coffee and tobacco are genuine auction crops, but cotton is not. There were no tea auctions in Africa in the 1930s, although tea is a typical auction crop. We will come back to these points later.

TABLE 12.1. Regular Agricultural Auctions in Africa Before World War II by Country, Town, Year Started, and Crop

Anglo-Egyptian Sudan	Tokar (circa 1932) cotton
	Port Sudan (1935) cotton
Kenya	Nairobi (1935) arabica coffee
	Mombasa (1937) robusta coffee
Southern Rhodesia	Salisbury (1936) tobacco
Nyasaland	Limbe (1938) tobacco
Northern Rhodesia	Fort Jameson (1939) tobacco

Sources: Gaitskell (1959, p. 188); Clements and Harben (1962, p. 120); McCracken (1983, p. 175); Haviland (1954, p. 378); Hill (1956, p. 99).

Note: Many colonial names have been changed. Anglo-Egyptian Sudan changed to Sudan; Southern Rhodesia to Zimbabwe; Nyasaland to Malawi; Northern Rhodesia to Zambia; Salisbury to Harare; and Fort Jameson to Chipata.

Views of Contemporary Observers

Early writings on the auctions in Africa are found in sources concerned with individual crops. However, there is no early review that examines all three types of auction together. It is nevertheless possible to find common elements in the views of contemporary writers. These writers welcomed the auctions as a commercial innovation because they considered them as favorable for the growers, whether for the coffee and tobacco growers (mainly Europeans) or for the cotton growers (mainly Africans). Support from the growers was said to be widespread. The colonial authorities, who shared the views of the growers, strengthened the new institutions by making sale at the auctions compulsory for all or most growers.

There is further a great deal of descriptive information in the early literature. Trade associations were formed with representatives from the ranks of both sellers and buyers. Auction rooms and halls were rented or built. Competent and impartial auctioneers were recruited. Rules about inspection and sampling procedures were agreed upon. The interval between successive auction days was fixed. Finally, agreements among growers about the scheduling and/or combining of supplies were worked out. As in other parts of the world, the excitement of auction bidding attracted observers and inspired them to write positive reports, often emphasizing the element of competition.

Implicit Views

By reading between the lines, it is possible to deduce from the early sources the basic convictions underlying the ongoing support for the auctions. First, the actors were convinced that differentiation was profitable. This conviction was not surprising in the case of tobacco. However, it was new in the case of cotton and coffee—crops for which the auction method brought a further differentiation, on top of the grades (see below). The belief in differentiation was probably strengthened by the economic crisis of the 1930s, which, like any other crisis, cast doubts on the system in existence at the time.[1] Second, the actors were convinced that, if supply was going to be differentiated, it should be done *at origin* and producers should be able to see the effect on prices at close hand. Third, the actors felt that the local auction market was a genuine compartment of the world market and that local prices were at par with the world prices. Fourth, the actors believed that the local classification of supplies was as good as that done in the world market and that local inspectors possessed the appropriate skills.

As we saw, all the new auctions were established in countries under British rule. This suggests that support for auctions may have drawn on a cultural factor, for the auction method had always been popular in Great Britain. The idea that physical markets should be open to the public was also firmly anchored in tradition. It is not surprising that proceedings at the auction markets in Africa vaunted the traditional aspects to impress the public.[2]

Some developments in transport also appear to have eased the way for auctions. First, conditions for travelers improved during the interwar period. It was therefore easier for importers in Europe to travel to East Africa and establish commercial contacts. By the mid-1920s, an informal coffee market had come into being in Mombasa (Stahl, 1951, p. 228). This, I feel, was a stepping stone for the auctions. Second, regular air services between East Africa and Great Britain were established in the early 1930s. (The London-Cape Town line, via Cairo, was inaugurated in 1932.) This permitted product samples to be sent by air to European importers, a practice which helped to harmonize the classification systems in the countries of origin and destination.

THE LOCAL AUCTIONS
FOR TRANS-OCEANIC CROPS

The auctions for cotton, tobacco, coffee, and tea, are briefly reviewed below, from their establishment until the present. This is followed by theoretical comments.

The Cotton Auctions

Local auctions for cotton were first tried out in the Anglo-Egyptian Sudan, where most of the country's cotton was grown on the Gezira Scheme, already discussed in Chapter 8. The Sudan Plantations Syndicate (SPS), which was responsible for the ginning and marketing of the Gezira cotton, began to sell at the local auctions in Port Sudan in 1936 (Gaitskill, 1959, p. 188). It did so reluctantly[3] because it was afraid that the prices obtained at the local auction would be lower than those obtained in Liverpool. However, the Sudan government put strong pressure on the company to use the auctions. The chief advantage seen by the government was that payment would be received much earlier than from sales in Liverpool. Early payment was highly valued in the depression years, not only by the government, but also by the Gezira tenants.

The decision of the SPS to sell at the auctions in Port Sudan was a compromise. Part of the Gezira cotton—particularly the higher quality cotton—continued to be sold in Liverpool, while the lower grades were auctioned in Port Sudan. The auctions were not very successful for the SPS, as bid prices frequently did not exceed the SPS reserve prices resulting in the withdrawal of much of their cotton.[4] The main groups of buyers at the auctions came from three large cotton centers: Alexandria (Egypt), Liverpool (Great Britain) and Bombay (India).[5] The SPS made samples available beforehand, but too late for them to be sent abroad. This favored buyers who had experts with full discretionary powers in Port Sudan, and militated against buyers whose representatives (employees or agents) had to consult their directors or principals and wait for confirmation by cable before making bids. The spinners, notably those in India, urged the SPS to release their samples earlier.

The Port Sudan auctions, suspended during the war, but resumed in 1947, served as a model for two other countries. In Uganda, cotton auctions were inaugurated in Kampala in March 1948, building on a less formally organized market of the 1930s (Thomas and Scott, 1935, p. 135). The cotton ginners in Uganda preferred local sales because this permitted inspection before purchase and thus facilitated marketing. In Tanganyika, cotton auctions were started in 1952. They were held in Dar es Salaam. At first, only cotton from the Eastern Province was sold there. In the 1957/58 season, cotton from the Lake Province in the country's northwest, until then sold in Kampala, was also channeled to the Dar es Salaam auctions (Yoshida, 1984, p. 94).

None of the cotton auctions remain in operation today. In Tanzania, the auctions were discontinued in the 1970s and replaced by tender sales or *tele-auctions*. The body responsible for the tender sales invites bids from a selected list of potential overseas buyers to whom it has first sent samples. The telexed replies are compared and the highest bidder is successful. Research on how and why the cotton auctions were ended seems desirable.

The Tobacco Auctions

The local tobacco auctions in the Rhodesias and Nyasaland were established between 1936 and 1939.[6] Most of the tobacco in those countries was grown by settler farmers. Among them were a few sophisticated entrepreneurs who understood the benefits of local auctions. In the 1930s they mustered general support within the sector, which gave the colonial authorities a basis to make selling at the auctions compulsory—a condition for sufficient turnover. It was further understood that only the more expensive tobacco, notably the flue-cured Virginia, could bear the additional costs of the auction system. For the burley tobacco, grown by African farmers and dried by simple air curing, it did not normally pay to auction the crop.

The auctions were modeled after the auctions in the United States, begun in Virginia in 1859; but while the American auctions have always been domestic affairs, those in Africa could not succeed unless they attracted overseas buyers. This may well explain why the tobacco auctions in the United States are numerous and small, while those in Africa are few and large. During the tobacco season, the auctions are held every day. All tobacco passes the auction floor, where the bales are inspected. Although the world tobacco market has long been dominated by an oligopoly of cigarette manufacturers, this has not prevented (either in the past or recently) a high degree of competition at the auctions.

The auctions continued during the war, albeit modified by wartime regulations. The auctions at Fort Jameson were discontinued in 1952, but after Zambia's independence, tobacco auctions were started in Lusaka in 1966. They were suspended in 1977 and have not resumed. In Malawi, both Virginia and burley tobacco are sold at the auctions in Limbe and Lilongwe. The latter auctions were established in 1979 to serve the tobacco growers in the northern and central regions of Malawi.

The auctions in Harare (formerly Salisbury) are claimed to be the largest tobacco auctions in the world. From 1965 to 1980, they were crippled by the international sanctions against Rhodesia. To permit sale by private treaty and thus circumvent the sanctions, the growers developed an elaborate grading system. It is significant that there was an immediate return to the auction system when the sanctions were lifted in 1980.

The Coffee Auctions

In 1935, auctions for arabica coffee were established in Nairobi, Kenya. The local trade association, the East African Mild Coffee Trade Association, wanted the auctions to serve the whole of British East Africa. The auctions indeed attracted supplies from Uganda and the Kilimanjaro area in northern Tanganyika. Most of the growers of arabica coffee were settler farmers, but there were also African farmers in Tanganyika.[7] The settler farmers (and estate owners) were the driving force behind the auctions. They realized that quality could be improved, not only on the farm, but also during subsequent processing. They established the Kenya Planters' Cooperative Union (KPCU), which built a modern coffee curing factory in Nairobi in 1937 and helped the *pooling of supplies* (Hill, 1956).[8]

The output of the KPCU mill was much better differentiated than that of the individual growers, which was not only a matter of better grading but also of superior classification. Although the cultivation of robusta coffee (mainly in Uganda and the Lake Victoria area of Tanganyika) was relatively unimportant, it was not forgotten. Auctions for robusta were established in Mombasa in 1937.

The Nairobi and Mombasa auctions, reopened in August 1947 (Leubuscher, 1956), served as models for new auctions in Kampala, Uganda, in 1953 and Moshi, Tanganyika in 1952. Those in Kampala competed with the auctions in Mombasa and undermined them. In 1964, the Kampala auctions were unexpectedly suspended, possibly for political reasons, but those in Moshi still function today. In the late 1980s, local auctions found renewed favor, notably among the donor organizations. For instance, the World Bank promoted new coffee auctions in Rwanda and Burundi. The World Bank also supported resumption of the local auctions in Uganda, but to no avail. Instead, a tender system was instituted in 1990. It may be argued that the cheaper tender system is suitable for the lower-priced robusta coffee (predominant in Uganda), whereas the auction system is appropriate for the higher-priced arabica coffee, as in Kenya.

In the 1950s, political considerations made it desirable to open coffee cultivation to African farmers in Kenya. This raised the question of whether these farmers could also sell at the auctions, and further whether they could do so under their own names. As the entry of a few thousand new sellers would have choked the auction system, it was necessary to exclude farmers with an output below a certain threshold. However, if these farmers formed a cooperative and merged their output, the cooperative was permitted to sell the joint output at the auctions. Most of these cooperatives have succeeded in keeping farmers interested in good quality.

The Tea Auctions

Local auctions for tea have been held in only two countries and were moreover established quite late: in Kenya in 1956 and in Malawi in 1969. The auctions in Kenya are the more important of the two. They were started in Nairobi, but moved to Mombasa in 1969. They have a certain popularity in neighboring countries, attracting considerable supplies from them (Houtkamp and Van der Laan, 1993). The trade associations that manage the tea auctions in Africa cannot ignore the continued strength of the tea auctions in London, still regarded as the "irreplaceable thermometer of price and quality of tea" (Forrest, 1985, p. 128). The competition between the central auctions in London and the local auctions in Africa is a fascinating subject. London auction officials claim that price formation at their auctions is superior, but the local auctions have a counter claim: in bringing forward the moment of sale, they provide financial certainty several weeks earlier. The managers of the London auctions, anxious about losing ground, devised the system of *off-shore auctions*, which has operated in conjunction with the regular auctions since 1982. Under this system, lots of tea from overseas can be sold on the basis of airmailed samples while the consignments to which they belong are still off shore, that is, in transit or about to be shipped.

Until about 1960, most tea growers in Africa were large plantation companies. Such large growers did not need the auctions, or at least not as much as the smaller producers. In fact, the tea plantation companies in Kenya and Malawi have adopted a policy of selling part of their output at the auctions and the remainder by private treaty—a policy approved by the government.

In Kenya and Malawi, tea cultivation by smallholders became important in the 1960s. Their entire production, after bulking and merging, is sold at the auctions. However, the merging of smallholders' supplies is not done via cooperatives, as with coffee in Kenya, but via a parastatal, the Kenya Tea Development Authority (KTDA). The KTDA buys fresh tea leaves from independent tea farmers for processing in its 40 factories. It has a reputation for high-quality "manufactured" tea and for harmonious relations with the farmers. In fact, many observers of the Kenyan tea sector feel that the KTDA arrangement is superior to that of the plantations where the work is done by wage laborers. The collaboration between the two actors—the tea farmers and the tea-processing KTDA—rests on the timely delivery of the fresh tea leaves (which may quickly spoil) and quick inspection and purchase by the KTDA. The unambiguous criterion for inspection (only the bud and the upper two leaves must be picked) appears to provide a sound basis for harmonious collaboration. The Smallholder Tea Authority in Malawi works on the same basis.

THEORETICAL CONSIDERATIONS

The world markets for tea, tobacco, and coffee recognize many classes, each with its own price. There is enough variation among these prices to make careful differentiation and classification rewarding. In Chapter 5, we stressed the role of the consumers in this fragmentation of demand (for instance, into niche and gourmet markets). Here we want to consider the role of the distributors with regard to demand.

By *distributor*, I mean a large actor who makes substantial investments in building a reputation with final consumers for consistent quality and taste. (The inherent variability of agricultural products does not make this an easy task.) Brand names, in which the reputation is embodied, play a significant role in the marketing of tea, tobacco, and coffee. (In my view, the cigarette manufacturers are primarily distributors.) The role of the importer-distributors in the tea, tobacco, and coffee TOMCs is a threefold one. First (and rather passively), importer-distributors monitor changes in consumer preferences and adjust their purchases. (They translate consumer demand into demand for the various production qualities distinguished in the world market.) Second, by means of their blending experience and their labels and brand names, importer-distributors standardize consumers' tastes. Third, by advertising, importer-distributors increase and stabilize global demand for tea, tobacco, and coffee. The growers of these crops, including those in Africa, benefit from this.[9]

In the countries of production, there has been a strong incentive for the tea, tobacco, and coffee growers to differentiate their supplies in line with the classification used in the world markets. This facilitates sales, but even more important is the fact that growers are unable to respond to world price signals unless their own classification approximates that used in the world market. This can be ensured by a corps of local inspectors whose skills equal those of the top inspectors in the world markets. In the tea sector, many local inspectors have been employees of the growers: the plantations were large enough to employ tea tasters.

Tea and Tobacco

Before the auctions were established in Africa, most of the tobacco seems to have been consigned to Europe for sale. Proper classification of consignments by the grower facilitated sales and brought higher prices. In the tea sector, both consignment sales and CIF trading took place. In the latter case, both exporters and importers were confident that the descriptions of the consignments were adequate and would not lead to litigation. Building and retaining a good reputation were therefore important for the tea plantation companies.[10] In the tea sector in East Africa, CIF trading proved adequate until the 1950s. Even after the auctions were established, CIF sales continued.

The advantages of selling at local auctions over CIF trading are much clearer for small growers. Since the auction method allows the buyer to inspect prior to purchase, the problem of disagreement on quality is practically ruled out. For smaller growers whose reputation is (still) uncertain, the local auction system removes many problems of breaking into the market. These problems are shifted to a new actor, the *buyer-exporter*. It is the latter who now has to build up a sound reputation abroad.

The local auctions brought a reshuffling of tasks and functions within the TOMC. We may illustrate this for the tobacco TOMC. Formerly, a tobacco grower was an exporter (although many tasks were delegated to a shipping and marketing agent). After local auctions were instituted, the grower ceased to be an exporter. This role was now assumed by a new buyer-exporter who bought tobacco at the auctions and subsequently exported it.

The functional reshuffling was accompanied by a change in the duration of the actors' *ownership interval*. Previously, the tobacco growers sold a consignment after shipment. Now they sold before shipment, which brought the moment of sale forward by several weeks. This reduced the ownership interval during which the growers had to finance their tobacco and were exposed to the price risk. In the light of the argument in Chapters 7 and 10, it is likely that the growers considered this interval reduction as a great benefit. This factor may also partially explain why local auctions flourished in East Africa, but were absent in West Africa. Due to the longer sea journey, the reduction in the ownership interval was greater for the growers in East Africa than it would have been for those in West Africa.

Some buyer-exporters (not only for tobacco, but also for the other auction crops) were employees of the importers. They were permanently (or seasonally) stationed in the towns where the auctions were held. Other buyer-exporters were independent businesspeople, trading on their own account or as agents or brokers for importers. Communication by cable and airlifted samples were crucial in the relationship between a principal and an agent. In the 1930s, the buyer-exporters' risks were primarily commercial. When the currency risk became important, notably in the 1970s, it was logical for this risk to be borne by the buyer-exporters.[11]

Coffee and Cotton

The introduction of local auctions for coffee and cotton in the 1930s marked a major commercial change. Products that had previously been considered as commodities and had been traded in bulk, were now differentiated and sold in much smaller lots. Moreover, this major change seemed to be a local, purely African phenomenon, not paralleled by developments elsewhere. Below, I offer three tentative explanations. Further research, whose scope includes producing countries outside Africa, will be necessary to determine which explanation is the most appropriate.

First, the conviction of the suppliers (coffee growers and cotton ginners) that auctions increased their net profits may have been illusory. In any case, it was more a matter of faith than of proof. As we saw in Chapter 5, it is practically impossible to present sound calculations. (I have never come across figures, however tentative.) It must further be pointed out that some of the advantages of the commodities were sacrificed. The across-the-board high marketability was given up. Generally speaking, while the standard qualities remained highly marketable, the special and the poor qualities had to be marketed with a far greater effort. Moreover, the moment-making freedom of the exporters was reduced because the sale of a particular lot cannot normally take place until it is available for inspection. The value of these two sacrifices cannot easily be expressed in monetary terms.

Second, it is possible that airmailed samples became a regular feature in commodity trading, changing the nature of the international trade in cotton and coffee during the 1930s. No doubt, these samples reduced an importer's risk compared to the previous situation in which a purchase was made on the basis of the grade only.[12] The SBS (sample before shipment) condition, which is now widely used in coffee contracts, points in the same direction. It gives an importer the right to cancel the contract if the quality of the airmailed sample is below standard. Although this right is seldom used, it helps to reduce an importer's risk. Here, the benefit of prior inspection is obtained without the complications of an auction system.

Third, a supplier's main objective in supporting the auctions may not have been to achieve greater differentiation in the product, but to gain the opportunity to sell locally. Local sales brought two advantages, notably for the smaller suppliers. The first was the suppliers' ability to sell their product, even if their international reputation was nonexistent or poor.[13] The second was a significant reduction in the ownership interval, similar to that for the tea and tobacco growers (see above). The suppliers needed finance for a shorter period and their exposure to the commercial risks ended sooner. Presumably, this was a major benefit in the crisis years when many growers and ginners were on the brink of bankruptcy. However, the reduction in the ownership interval proved to be smaller than expected. At first suppliers were given a free hand in determining the moment of auctioning their product. Unfortunately, this made the physical flow at the auctions very irregular, which was unattractive for the buyers. This was solved by *orderly marketing* arrangements, that is, the scheduling of supplies according to a strict, preset timetable.

The Future

Auctions are presently in great favor. They are recommended because of the benefits they ensure: a high degree of competition and price transparency. Unfortunately, many of the points raised in this chapter are not taken on board in these arguments.

With Cassady (1967), I hold the view that regular auctions are most appropriate for products with "subtle quality characteristics." Since only tea, tobacco, and coffee have such characteristics, there is no point in recommending the auction method for other trans-oceanic crops. In addition, I want to point out that auction systems are vulnerable. Minimum conditions of logistic management, honesty, political stability, and reliable telecommunications have to be fulfilled to keep an auction system healthy and attractive.

Glossary

at origin: At the beginning of the marketing channel. The term refers to the site of production or, more loosely, the district or country of production.

buyer-exporter: Here defined as an actor in a TOMC who is a buyer at the local auctions as well as an exporter.

distributor: An entrepreneur who has made substantial investments in building a reputation for consistent quality and taste. The entrepreneur's reputation is usually embodied in a brand name which is known to many final consumers.

off-shore auctions: Auctions at which lots of, for instance, tea that have not yet arrived in the country of destination are sold on the basis of airmailed samples.

orderly marketing [at local auctions]: Arrangements which spread out supplies over time by means of strict delivery timetables for participating growers.

ownership interval: Here defined as the interval during which the grower or processor owns the product, is responsible for financing, and exposed to various risks.

pooling of supplies: Action which takes place when a group of growers instruct a central organization to merge (or pool) all or part of their supplies and to sell the pooled supplies under its name.

tele-auction: A special form of auction where the bidding is done in writing, for instance, in a telex message.

Chapter 13

Perishable Crops
and Their TOMCs

The physical risks of rapid spoiling and deterioration of quality put perishable crops in a special category and consequently require a special organization of the TOMC. For our purposes, a perishable crop is one which is perishable at the point of shipment.[1]

Only two crops need to be discussed in this chapter: bananas and pineapples. This limited number may come as a surprise to readers in Europe who have seen many other fresh fruits and vegetables from Africa in local supermarkets. However, all this fresh produce comes by air and, as it does not pass through a TOMC, is omitted from our discussion.[2] It is further relevant that the statistics often add the word "fresh" to distinguish fresh bananas from dried ones (which were popular in the past), and fresh pineapple from the canned fruit. Most of this chapter deals with bananas.

OCEAN TRANSPORT: REFRIGERATED SHIPS

For crops that are perishable when offered for shipment, special refrigerated ships are necessary. Such ships were first developed to carry deep-frozen meat from Argentina, Australia, etc., to Europe. By 1890, such ships were carrying significant volumes. Around 1900, techniques were improved in order to carry fruit, whose transport also requires low temperatures. Appropriately equipped ships were first introduced to carry bananas from Jamaica to Great Britain. It was only around 1930 that such ships were used for African bananas.

Having refrigerated ships running on fixed time schedules between a port of shipment and a port of destination is only profitable if certain conditions are fulfilled. First, cultivation must take place close to the port so as to reduce local transport to a minimum. Second, harvesting must be scheduled in such a way that a ship can be rapidly and completely loaded as soon as it arrives. Third, there must be unloading facilities and cooling and/or ripening installations in the port of destination. Fourth, transport after shipment must be both coordinated and adequate: refrigerated trains or lorries are needed to carry the fruit to the wholesalers. Management always has to be aware of the fatal results from delays, stoppages, and bottlenecks.

Compared with the durable crops discussed in earlier chapters, there are further differences in that the storage periods should be as short as possible; the (pipeline) stocks are consequently negligible. In addition, negotiating the sale and export of 500 tons of a perishable product is a very different proposition from selling a similar quantity of a durable product. In general, the selling of perishables does not take place consignment by consignment, but within a contractual framework covering a period of one or more years, where the size and timing of individual consignments are specified within narrow limits.

BANANAS

The Early Projects

In the interwar period, businesspeople of three countries on the European continent (Germany, Italy, and France) initiated banana growing and exporting projects in Africa. It is significant that there was no similar British project. Britain had been importing bananas first from the Canary Islands and since 1900 from Jamaica. In fact, the banana trade was well established in Britain. It was dominated by Elders & Fyffes, which since 1913 had been part of the United Fruit Company (UFC), a large banana company operating in the Western Hemisphere. Elders & Fyffes had branched out to the European continent and had established subsidiaries in Germany and France, where consumption per capita had risen considerably, partly due to vigorous advertising.

The information presented here on the projects of the 1930s is taken from Houtkamp, 1996. In this book the German, Italian, and French projects are described and compared.[3]

The German project rested on the cooperation between a large fruit importer in Hamburg and a shipping company. When considering suitable locations for banana plantations in Africa, they chose the southeastern part of Nigeria, an area which had been a part of the German colony of Kamerun prior to World War I. (After the war, Kamerun had been divided between France and Britain as mandatory powers.) German businesspeople had owned plantations in this area during the German period, and although these plantations had been confiscated during the war, most of them had been repurchased by Germans in the second half of the 1920s. It was on these plantations that the cultivation of bananas began. Soon, German ships were running regularly between Tiko and Hamburg. The Nigerian authorities were pleased with the additional exports.

Italian planters established banana plantations in Italian Somaliland in the late 1920s. At first, the planters sent their bananas with general cargo ships equipped with one refrigerated hold. Later, proper banana carriers were used. The Italian venture had to cope with two handicaps: production was expensive because of irrigation, which was essential, and demand in Italy was limited because per capita consumption was still low. In 1935, the Italian government, concerned about the precarious economic position of Somaliland, stepped in and created a state monopoly which brought about far-reaching coordination between the planters, the shipping companies, and the fruit distributors in Italy.

Planters in French Guinea had long tried to grow bananas for the French market. Of all the French colonies in Tropical Africa, Guinea was nearest to France. However, without refrigerated ships, success remained uncertain. Help from fruit importers in France was uncertain because they could also buy bananas from Martinique and Guadeloupe in the Caribbean. The prospects of the planters in Guinea improved in 1932 when the French government began to bar foreign products from the French market and to give privileged access to colonial products. These protective measures also made the banana trade more attractive to the shipping lines, which now invested in banana carriers. Apart from French Guinea, Côte d'Ivoire and French Cameroon began to grow and export bananas.[4] It is interesting that UFC established its own banana plantations in Cameroon in 1935. UFC had suffered from the French measures which excluded bananas from its plantations in the Western hemisphere. However, UFC bananas, grown under the French flag in Cameroon, could not be barred from the French market.

What were the common features of these banana projects? First, all of them faced a high threshold because in the early 1930s, only relatively big banana carriers were economical. Unless a new project started on a large scale, corresponding to the capacity of three or more carriers operating in a regular schedule, it could not succeed. Second, government protection with import licenses, quotas, or tariffs was important. Third, the new TOMCs displayed a high level of coordination of activities in all three sections: upper, middle, and lower. This was essential because of the risk that the bananas would spoil before reaching the consumers. In view of the evident political aspect of the TOMCs (the matching of colony and metropolitan power which occurred), it seems likely that political support (both administrative and economic) strengthened the coordination. This raises the question of whether the African banana TOMCs, in spite of their rapid success in the 1930s, were viable in themselves.

World War II and the Postwar Period

The banana projects of the 1930s had hardly overcome their teething problems when the war broke out and disrupted trans-oceanic transport. When Italian Somaliland was occupied by Allied forces in 1941, exports ceased. French West Africa was largely cut off from France because of the Allied blockade and all refrigerated ships were withdrawn. In Nigeria the banana sector was taken over by Elders & Fyffes almost immediately after the outbreak of hostilities. Shipments continued, but now destined to Great Britain instead of Germany (Davies, 1990). This was only a temporary arrangement because in 1941, the British government banned all banana importing in order to save shipping space for more strategic products. After the war, exports were resumed, again destined to Great Britain. One could argue that a British banana project accidentally emerged in Africa, simply because Britain inherited the German project.

In the early postwar period, European demand was strong as consumers sought to make up for the years without bananas. All prewar projects were revived, but no new TOMCs were created. The role of large companies increased. UFC came to Somaliland, then a trust territory under the United Nations, and took over the Italian project. In Nigeria, Elders & Fyffes now operated on a permanent basis; and UFC resumed its activities in French Cameroon.

The end of the colonial era brought several changes. First, the region in Nigeria where the banana plantations were located became part of a reunified Cameroon and entered the Franc zone. What had been two banana sectors, one in Nigeria and one in Cameroon, became one sector, dominated by UFC. Second, a drastic change occurred in French Guinea. Soon after this country decided by referendum to leave the French community of states, its banana TOMC began to disintegrate, showing that without political support, it could not survive. Côte d'Ivoire benefited from Guinea's decline because its bananas replaced those from Guinea.

In the 1960s, the lack of viability which I hinted at began to manifest itself. In Cameroon, all bananas were directed to France. This marked the beginning of a long period of decline, culminating in the withdrawal of United Brands, the successor of UFC. After a period of parastatal control, the Del Monte Company came to Cameroon later in the decade and assumed responsibility for some plantations. The banana TOMC between Somaliland and Italy continued in good shape after the former gained independence in 1960. It survived even in periods when the Suez Canal was closed—a closure which forced the banana carriers to travel around the Cape. However, in the late 1980s, the banana sector collapsed when civil war brought strife and disruption to southern Somalia. The country that achieved the best results has been Côte d'Ivoire. Its prospects further improved in 1993 when the European Union adopted a common import policy for bananas, favoring the ACP countries.

Trans-National Corporations

The banana TOMCs are made rigid by the strict sailing schedules for the ships. In such circumstances, it is difficult for the businesspeople involved to rely on the market mechanism for the coordination of the TOMC. Instead, businesspeople create a large organization which brings exporters and importers under the same management. This is what I called a vertical ocean-straddling marketing system (VOSMS), and the companies that create and operate such a system belong to the category of the *trans-national corporations* (TNC). This term came into vogue in the 1970s when sensitivity to foreign influence in newly independent countries was high. However, the phenomenon is much older. Indeed, one may argue that the colonial situation favored the creation of ocean-straddling TNCs.[5]

The global banana sector has long been dominated by TNCs. The big three TNCs are United Brands (formerly UFC), Del Monte, and Dole. The first two have operated in Africa, as mentioned. Because of the economic power of the banana TOMCs, African governments tend to be wary of them. They suspect a hidden inequity in the transfer prices which the TNCs use when exporting bananas. However, it is inherent in the nature of ocean-straddling TNCs that the transfer prices cannot be evaluated; it is equally hard to prove that they are fair or unfair. A yardstick for price comparison is lacking. It is not my intention to deny the proven and suspected drawbacks of these and similar TNCs, but their advantages for Africa are significant: TNCs solve the problems of trans-oceanic trade in perishable crops, and have invested enough money in the TOMC, on either side of the ocean, to be a guarantor of continuity in the interests of African governments and TNC employees.

FRESH SEABORNE PINEAPPLES

We will close this chapter with a short paragraph on seaborne fresh pineapples exported from Côte d'Ivoire. (For canned pineapple, see Chapter 11.) Since the same type of refrigerated ship is used for pineapples and bananas, it was easy for exporters in Côte d'Ivoire to experiment with the export of fresh pineapples. This began in the 1950s. Like the bananas, the pineapples were mainly shipped to France.

It is a special feature of Côte d'Ivoire that it exports both canned and fresh pineapple. Sawadogo (1977) emphasizes that the two subsectors differ in cultivation practices and are geographically separated. The relative importance of the two subsectors, measured by export value, has varied over the years.

Glossary

trans-national corporation (TNC): Literally, a company or corporation that operates in two or more countries. The term is often used in a special context: the company headquarters are located in a rich, technologically advanced country and the company subsidiaries in poor countries.

PART III:
NEW FINDINGS
OF THE TOMC APPROACH

Chapter 14

Implications for the Export Basket and Agricultural Policy

In the preceding seven chapters, the TOMCs for Tropical Africa's export crops were intensively examined, and, although an initial examination may miss certain points, valuable insights have been gained. First, there are considerable organizational differences between the TOMCs for (a) commodities, (b) auction crops, (c) minor crops, and (d) perishable crops. Second, high marketability is an important product characteristic. Third, the TOMCs for highly marketable crops are easier to organize and to keep going. Fourth, there is a close relationship between high marketability and the existence of an institutional world market.[1]

If I were to observe the self-imposed restrictions of this book (only Tropical Africa, only agriculture), my sole task would be to set out the implications of these insights for African ministers of agriculture. However, I have come to the conclusion that my findings have a wider relevance. They have implications for most less developed countries (LDCs), notably for those whose export basket mainly consists of trans-oceanic products.

The *export basket (portfolio) of a country* is a list of all export products specifying the volume and value of each product. The total basket may be subdivided into sectoral baskets (such as the agricultural one of this book). The composition of the export basket is rarely constant, but changes over time. Governments may approve of these changes or consider them undesirable. In the latter case, governments may actively influence the composition.

When we consider the whole export basket, we can benefit from what scholars of international trade and development economics have written about its actual, potential, or ideal composition. An undercurrent in these writings was the view that the basket of the LDCs was far from ideal. This prompted me to develop the theory of *exporter preference.*

The presentation in this chapter involves four steps: (1) a consideration of theories and facts about the export basket and export diversification, (2) the presentation of the theory of exporter preference, (3) a reinterpretation of Tropical Africa's agricultural export record in the light of this theory, and (4) the formulation of recommendations to African governments.

THEORIES AND FACTS ABOUT THE EXPORT BASKET

We first review what some well-known theories say about the ideal export basket and how it comes about. The classical theory of comparative advantage/costs developed by Ricardo (1817) concludes that the selection of export products is done by producers on the basis of low comparative costs. These costs determine which products can be profitably exchanged. Since producers have the best position to compare potential export products, they should be given a free hand to select the most promising ones. The basket that spontaneously comes about as the result of the decisions of many producers ensures an ideal allocation of resources on a global level and a pattern of international trade that is profitable for all participants. The theory of factor endowment (Ohlin, 1933) comes to a similar conclusion: export products are selected on the basis of the high comparative abundance of one of the factors of production: land, labor, or capital.[2] Both theories may be said to rest on producer decisions or on what I would like to call *producer preference*—a term to be used later.

There have always been vague doubts about the classical conclusions. In 1950, they were taken up by two economists, Prebisch and Singer. They argued that producers of primary commodities benefited less from international trade than manufacturers. When the long-run terms of trade between primary and industrial products were studied, the gains were smaller for the primary producers. Differences in the process of price formation for the two categories were considered to be responsible for this. Since primary products are mainly exported by the LDCs and industrial products mainly by the developed countries, the economists' conclusion was that the LDCs benefited less from international trade than the developed countries. The policy implications were clear. The governments of the primary commodity exporting countries should modernize their economies. In particular, they should promote industrialization, and favor industrial rather than agricultural or mineral exports. By altering its export basket, a country could advance in international trade to a more advantageous position.

Another economist who had doubts about the existing export baskets of the LDCs was Arthur Lewis. He attributed the low prices for tropical primary commodities to the low productivity of tropical farmers. In countries where an export crop was the only alternative for subsistence farming, farmers were willing to grow this export crop, even if prices were very low. Lewis believed that the terms of trade for tropical products could only improve if the productivity of tropical agriculture (for export and subsistence crops) increased.[3] He did not recommend direct intervention to change the export basket.

The views of Prebisch, Singer, and Lewis appealed to those who had observed that the export basket of many LDCs had been characterized by its narrowness. Only a few products accounted for the bulk of the export earnings. This state of affairs made a country vulnerable. If demand for one of these products fell, the whole economy was dragged down. Moreover, several of these products were known for strong price fluctuations, which increased the variability of a country's export earnings.

These concerns have always made sense to me. The composition of the export basket of most African countries seemed lopsided and far from ideal, but I believed that the real cause had not been discovered. Before I give my own interpretation of how export products are selected, we must consider some elementary facts about the export basket.

To describe an export basket, one has to make use of the foreign trade statistics which are generally based on the Standard International Trade Classification (SITC). SITC statistics facilitate comparison between countries, but suppress some detail. Moreover, unimportant products are combined into categories, some of which are heterogeneous, as the adjective "miscellaneous," indicates. As a result merely counting the individual export products is an unsatisfactory approach for measuring a country's export basket. Two other measures have therefore been adopted: *concentration* and *dependence*. In the former, one asks how many products account for 75 percent of the country's export earnings. In the latter, one presents the percentage of the export earnings that is earned by the one, two, or three most important products. These two measures are commonly used, for instance, by UNC-TAD, but we should be aware of the measures' fluidity due to the variability of export earnings. Table 14.1 is based on the concentration measure for 40 countries in Tropical Africa. In this connection, we should also mention *export diversification*. Export diversification[4] is basically a process by which a country's export basket widens—a process that normally accompanies economic development. When this process occurs too slowly or not at all, governments tend to adopt export diversification policies.

THE THEORY OF EXPORTER PREFERENCE

I have developed a new theory to explain why an actual export basket deviates from the ideal one of the classical theories. The theory is valid in the following circumstances. First, the exporter and the producer, while operating in one marketing channel, have to be separate actors, not only economically, but also culturally. A cultural difference exists, for instance, when the producer is rather traditional and the exporter more modern. Such a difference ham-

TABLE 14.1. Tropical African Countries by Degree of Export Concentration, 1972-1976 and 1982-1986

75% of Export Earnings Accounted for by	Number of Countries 1972-1976	1982-1986
one product	6	13
two products	17	9
three products	6	4
four or more products	11	14
	40	40

Adapted from Barratt Brown and Tiffen (1992), pp. 22 and 165-169.

pers mutual understanding and collaboration in the marketing channel. Second, exporter preference must differ from producer preference, that is, the products which the exporter prefers to export are not the same as those the producer prefers. The new theory contends that in these circumstances, the composition of the export basket becomes distorted whenever exporter preference overrules producer preference. Three points are involved: the nature of exporter preference, the weakness of the producers, and the scarcity of exporting competence.

The Preference for Highly Marketable Products

One of the conclusions of the preceding chapters has been that exporters have a strong preference for highly marketable products. I have found five reasons for this preference, all of them appearing in the context of agricultural products. First, market intelligence for current and future prices is served up on a silver platter. Second, many practical marketing problems are removed or, put differently, the transaction costs are low. Third, the market offers short-run continuity, as it is easy for an exporter to find a buyer at short notice. Fourth, the market offers long-run continuity because it is acceptable for an exporter to switch from one buyer to another. Fifth, the transparency of the market encourages supporting enterprises such as shipping and insurance companies, and commercial banks to provide extensive services in the area of transport, physical-risk-bearing, and short-term finance. My hypothesis is that the same reasons are valid for the exporters of highly marketable nonagricultural products. Whenever exporters have their way, they increase the share of the highly marketable products in the export basket.

Weakness of the Producer

Exporter preference operates most strongly when the producer is small in relation to the exporter. A situation in which large exporters collaborate with small producers has been common in many LDCs, notably in agriculture, but also in the surface mining of gold and diamonds, the gathering of forest products, and in arts and crafts. However, the producers' weakness is not only a matter of scale. It also depends on their ability to defend their interests when exporter preference differs from and conflicts with producer preference. When the exporter selects a product which is economically unattractive for the producers, the latter may refuse to cooperate or withdraw. Withdrawal is the main instrument of countervailing power: it cancels the exporter's power partly or completely.[5] The question must now be asked whether for one reason or another, the producers are reluctant to withdraw. This is the case when product A, chosen by an exporter, has a large *quasi rent* component, that is, alternative production opportunities are much less profitable and the producers are unwilling to switch to them.[6] When the producers are unwilling to withdraw from the production of A, they are weak and the exporter is strong.

We may elaborate this as follows: suppose a situation of global oversupply for primary product A emerges due to high exports from several countries as a result of exporter preference for A. Do the falling world prices for A discourage the exporters and cause a reduction of exports? This is the normal reaction as envisaged by classical economics. However, in this case, it may not work for the following reasons. First, high marketability, which depends on the way the market is organized, is not reduced by falling demand and continues to attract exporters. Second, exporter profits do not depend on prices, but on trading margins. If the exporters are able to reduce their buying price far enough to maintain their margin, there is no reason to abandon product A. The exporters are, in fact, able to reduce their buying price when the producer does not withdraw, but continues to produce A in spite of the lower price.

The Scarcity of Exporting Competence

Each country has what I would call a *fund of exporting competence*. The fund consists of at least three components: exporting skills, financial resources, and enterprises with an international reputation. This fund is relevant because if it is small, the effects of exporter preference tend to be strong. To explain this, I must introduce the concept of *exporter reluctance*—the counterpart of exporter preference. Indeed, exporters are reluctant to export products that are difficult to market, that is, where the obstacles to export marketing are high. In a country where the fund of exporting competence is small, exporters tend to concentrate on the highly marketable products and shun the difficult-to-market products.[7] Exporters make no effort to handle these products, let alone to promote them. A small fund of exporting competence tends to breed complacent exporters who seek the way of least resistance.[8]

Three Facts Explained and General Recommendations

The new theory accounts for three facts for which until now no satisfactory explanation was available. First, the preponderance of primary commodities in many export baskets is now seen as the result of the fact that most primary commodities are highly marketable. Second, the narrowness of many export baskets is the result of the fact that only a few products are highly marketable. (Although few in number, they represent large volumes.) Third, the variability of the export earnings of the countries with narrow export baskets is the result of the fact that price fluctuations are common for highly marketable products. As I see it, price variability and high marketability are two sides of the same coin.

The prevalence of exporter reluctance further explains why the export basket of countries with a small fund of exporting competence is easily subject to a process of *erosion*, a term I coin to indicate the opposite of diversification. Exporter reluctance often causes these countries to slide back to a basket exclusively composed of highly marketable products. If this happens, the uphill task of diversification has to start all over again. This has presumably been one of the reasons why there has been so much disillusionment with diversification policies in the past.[9]

When trans-oceanic trade expanded rapidly in the later decades of the nineteenth century, the fund of exporting competence was small in many LDCs. The gap was partly filled by foreign, mainly European, businesspeople. Opinions about them have been ambivalent. On the one hand, their role and competence were essential elements in the growth of primary exports; but on the other hand, particularly in countries where they were few in number, these exporters were largely responsible for the narrow *primary commodity orientation of a country*. They may have obstructed or retarded the diversification of exports because of their exporter reluctance, although this was not clearly perceived at the time.

Among the countries to which the new theory applies, we must distinguish two categories: in one, exporter preference is practically unopposed, but in the other, it is strongly counteracted by producer preference. In countries of the second category, a mixed situation exists in which both producers and exporters have a significant effect on the composition of the export basket. The conventional concept of the export sector is now seen to be misleading. The sector is not a policy unit, but the result of conflicting goals and policies: not only producer preference, but also exporter preference and reluctance play a role.

The new theory leads to a few broad recommendations to the governments of LDCs that are worried about the narrowness of the export basket of their country. First, strong producers are important because they limit the power of the exporters and counteract the exporters' preference for highly marketable products. The policy prescription is clear. Part of a successful diversification policy is the encouragement of strong producers in sectors that have until now been dominated by exporters. Even if the majority of the producers remains weak, a few strong ones may improve the situation. Second, the fund of exporting competence should be increased. This can be done by training and by recruitment from abroad. Since measures, such as nationalization or privatization of the exporting function, have in principle no effect on the size of the fund, it is illusory to expect miracles from either of them. Finally, if the fund of competence remains small, the allocation of the skills available assumes special importance. The dominant view has been that such skills should be used to market industrial products. However, the returns on other products may be higher.

REINTERPRETING AFRICA'S AGRICULTURAL EXPORT RECORD

The theory presented above is the result of what I learned about Africa. Below, I present and interpret the evidence I have found. Fortunately, there is a wealth of information on the trans-oceanic exports of Tropical Africa, as seaborne exports and imports have been reliably recorded for a long time. Most comprehensive studies cover all export products (Neumark, 1964; Mitchell, 1982; World Bank, 1989; and Wrigley, 1990), but we will confine ourselves to the agricultural

ones. The analysis below is conducted in terms of the crop classification developed in Chapter 5 and used in subsequent chapters.

We will first look at the agricultural export basket of Tropical
Africa as a whole. Recall that many commodities and minor crops
have a long history. They were already exported in the period before
1880. The auction crops began to appear just before World War I.
The exports of perishable crops came last, that is, in the 1930s. By
this standard, it is clear that Tropical Africa's export basket widened
over time.

Interesting points emerge when we differentiate the historical
record on a geographical basis and contrast the agricultural export
basket of Atlantic Africa with that of Indian Ocean Africa. In this
exercise, it is helpful to distinguish three periods: before World War
I, between the wars, and after World War II. In the first period,
Atlantic Africa, and in particular West Africa, had a head start with
regard to all exports, mainly because of its proximity to European
markets. Agricultural exports were much higher than in Indian
Ocean Africa, both in absolute terms and in relation to population. In
Atlantic Africa, the export function was almost entirely in the hands
of European companies who helped this part of Africa to "plug into"
the modernizing world markets for oilseeds and vegetable oil.

The Interwar Period

The interwar period has often been seen as one of stagnation and
complacency. For Atlantic Africa, this was largely true, although
the inauguration of cotton cultivation in French Equatorial Africa
and the Belgian Congo should not be overlooked. By contrast, it
was during this period that Indian Ocean Africa made good progress and began to catch up. One line of advance was formed by the
auction crops, which became significant in this region, but
remained practically absent in Atlantic Africa.[10] Since the growing
of auction crops places a stronger emphasis on quality than the
other crops, it requires knowledgeable and sophisticated growers,
growers who were more numerous on the eastern side of the continent. Sophisticated growers help to shape an export basket in accordance with producer preference. That is obvious when they operate
as grower-exporters, but it is also true when they are merely grow-

ers, and even when the trader-exporters, on whom the growers depend, are reluctant to handle the crop initiated by them.

. The settler farmers in British colonial Africa are the best-known examples of sophisticated growers. Settler farmers became the target of a great deal of critical literature in the 1970s (Brett, 1973; Van Zwanenberg, 1975; Palmer and Parsons, 1977). At issue was their use of political influence to obtain economic privileges, often at the expense of African farmers. This charge may well be justified, but it should not eclipse the point I want to make here. The settler farmers counteracted exporter preference. They persuaded exporters to handle new crops and so helped diversify Africa's agricultural exports. Moreover, they increased the fund of exporting competence, as is demonstrated by their introduction of commercial innovations such as the local auction system. Entrepreneurs similar to the settler farmers operated in the French, Belgian, Portuguese, and Italian colonies.[11] A number of them were important in the inauguration of banana exports in the 1930s.

Since settler farmers were virtually absent in West Africa, nearly all growers of export crops were Africans. The territories in this region therefore came to be described as peasant economies.[12] With equal right, they might have been called *foreign exporter economies*. Most of the time, these exporters were in a position to pass a lower world price on to the growers because the latter did not easily withdraw from production, even if prices dropped to very low levels. This explains the strength of the foreign exporters and their unchallenged position as channel leaders. In the 1930s, their power was analyzed in terms of monopoly (later refined to oligopoly), but the concept of countervailing power provides a more appropriate tool. The almost complete absence of this power among the growers was responsible for the skewed power distribution in the channel.

It appears that for most of the interwar period, the foreign exporters in West Africa were complacent. Little attention was given to examining the export potential of new crops. The fund of exporting competence remained small. These factors explain why the process of gravitation toward the commodities and the narrow commodity orientation were so strong in West Africa.[13]

The Period After World War II

The main event after World War II was the establishment of public exporters (EMBs and *Caisses de Stabilisation*). Neither the colonial authorities nor the African governments used them to counteract exporter preference. On the contrary, I am inclined to say that the EMBs fossilized the narrow commodity orientation of Atlantic Africa. Presumably, the African governments assumed that public exporters were ready allies in their diversification policies. However, this ignores the reluctance factor. Exporters may actually obstruct the government's diversification efforts. In many African countries, the EMBs did little to diversify agricultural exports, even in countries where the ministry of agriculture attempted to do so. Any explanation of this apparent contradiction must begin by recognizing that the EMBs were highly specialized organizations and retained a large degree of independence. Furthermore, most of them were not well suited for the export of the difficult-to-market crops. My impression is that most African EMBs performed poorly in selling these crops and that their support for diversification was lukewarm.[14]

The foreign element in agriculture and trade became smaller and weaker after independence. In West Africa, this occurred when the foreign trading companies were replaced by national EMBs. Elsewhere, notably in Indian Ocean Africa, the role of the settler farmers declined. My hypothesis is that the decline of the foreign element reduced the number of sophisticated growers and the fund of exporting competence and, indirectly, led to the erosion of the export baskets. However, it is not easy to assemble data to verify this hypothesis.[15] In this connection, we should remember that the commercial institutions which the settler farmers had created remained intact and proved useful to the African growers who took their place in, for instance, the tea and coffee sectors. As a result, the export basket of Indian Ocean Africa today remains wider than that of Atlantic Africa.

A new foreign element appeared in the 1960s and 1970s, when a number of agricultural trans-national corporations turned their attention to Africa. Most of them focused on perishable fruits, which they exported in fresh or preserved form. They helped diversify Africa's exports.

RECOMMENDATIONS TO AFRICAN GOVERNMENTS

The analysis above has certain implications for African governments. If they are worried about a narrow export basket for agricultural crops, the general recommendations mentioned earlier are relevant: the government should strengthen producers and increase the fund of exporting competence. The implications for the minister of agriculture are more specific. The agricultural export sector is too large and unwieldy for it to be dealt with as a single entity. Some kind of division of the sector is necessary. One division is that between large-scale grower-exporters who can fend for themselves, and small farmers who may need the government as a protector or an arbiter. For small-scale export agriculture, I recommend the division into four categories on which this book is based. It is not the only division that is possible or useful, but it is one that provides new insights. Below, I review the pros and cons of the subsectors corresponding to these four categories. This review is helpful for the ministry not only when diversification policies are discussed, but also when the whole sector experiences a crisis and its survival is at stake. Should certain subsectors be helped above others? Or, if no funds are available, what advice should be given? The issue of continuity receives special attention. As we saw, MCA attaches great value to the continuity of the channel. Our perceptions on this point have therefore been sharpened.

The subsector of the commodities has a poor economic reputation due to the great price fluctuations associated with these crops. Suggestions to expand the production of a commodity or to start with a new crop of this kind are bound to be opposed. However, the commodities have a high rating for continuity—an advantage which is often overlooked. A commodity subsector may experience a serious downswing, but a total collapse is unlikely. Indeed, the well-organized world markets not only reflect the continuing significance of these crops, but also foster continuity. LDCs producing these crops are sure of a permanent outlet.

The subsector of the perishable crops has a poor political reputation. The great drawback is the presence of TNCs, which may bring a large degree of foreign control over the subsector. However, the advantage is that the TNCs offer continuity: they do not invest unless they intend to stay. Governments need a framework to simultaneously evaluate continuity (which they want) and foreign control (which they dislike).

The subsector of the auction crops has always had a good reputation. In recent years, it has been strongly recommended by donor organizations because of the transparency and integrity of the auction selling method. I have shown the other side of the coin in Chapter 12: the auction system is complex; it is therefore vulnerable and costly. Special skills are necessary and the average marketing costs are much higher than for the commodities. This recognition may not come easily in a country whose experience has been limited to the commodities. An often overlooked advantage of the auction crops is that many importer-distributors spend large sums on advertising. This helps to increase and stabilize global demand for these crops.

It is difficult to generalize about the subsector of the minor crops because these crops form a residual category. Some crops resemble the commodities in the sense that the continuity lies in the market. Others resemble the perishable crops because a permanent relationship with one or a few importers is essential; a rupture in the relationship may lead to the total collapse of cultivation. This illustrates a general point: the continuity of a subsector in Africa rests either on the impersonal forces of the market or on personal relations with particular organizations and people on the other side of the ocean. By putting it this way the subjective element is highlighted. Whether or not a decision maker in Africa prefers to be dependent on people or on impersonal forces depends very much on subjective considerations. There seems to me a real danger that support for (or withdrawal of support from) a subsector will be determined by one-sided, subjective considerations.

These are by no means the only relevant points in agricultural policies; I have raised them because they are rarely mentioned in the economic discussions.

Glossary

concentration [of products in the export basket]: A measure which indicates how many main products account for, say, 75 percent of a country's export earnings.

dependence [on main export products]: A measure which indicates the percentage of a country's export earnings that is earned by the one, two, or three most important products.

erosion [of the export basket]: A process by which a country's export basket becomes narrower, that is, contains fewer products than before.

export basket (portfolio) of a country: A list of all export products, specifying the volume and value of each product. The products on the list are usually ranked by value.

export diversification: A process by which a country's export basket becomes wider, that is, contains more products than before.

exporter preference: The exporter's wish and/or decision to export particular products.

exporter reluctance: The exporter's wish and/or decision not to export particular products.

foreign exporter economies: Economies whose production and export structure is determined by foreign exporters rather than by local producers.

fund of exporting competence: The total of exporting competence in a country. A small fund may hold back exports or may lead to a concentration on easy-to-export products.

primary commodity orientation of a country: A process in which the share of primary commodities in the export basket increases.

producer preference: The producer's wish and/or decision to export particular products (or to have them exported by an independent exporter).

quasi rent: A temporary profit found in a sector whose earnings are in excess of the opportunity costs of one or more of the factors of production employed in the sector.

Chapter 15

Conclusion

The adoption of the TOMC as an analytical tool has opened up a new path for studying African export agriculture. The exercise has borne practical and theoretical fruits. On the practical side, I mention renewed interest in subjects that have been ignored for many years such as the telegraph, the auction system, and the banana exports (Chapters 4, 12, and 13). Another practical fruit took the form of new interpretations of particular events, problems, and relationships (Chapters 8 and 9). On the theoretical side new insights were gained with regard to the classification of crops (Chapter 5), the selling conduct of the exporter (Chapter 7), the velocity of the product flow (Chapter 10), and the exporters' preference for highly marketable products (Chapter 14).

It is now time to put the new theoretical findings in a broader context. I have selected three of these: (1) academic studies, (2) Fair Trade organizations, and (3) African governments.

IMPLICATIONS FOR ACADEMIC STUDIES

The long-established fields of study that have generated pertinent theory are: (1) agricultural marketing, (2) international trade, (3) development economics, and (4) economic history. In considering each field in turn, the first task will be to sketch the overlap between my findings and the theories—an overlap which is substantial in some fields.

Agricultural Marketing

To indicate the area of overlap it is helpful to use the model of an hourglass (Kohls and Uhl,1990, p. 27). Many agricultural systems resemble an hourglass, wide at the top (many farmers) and at the bottom (many consumers). In domestic systems, the entire hour-glass is located in one country and the economic conditions at the top and the bottom are more or less the same. This is not the case in trans-oceanic systems. In fact, the conditions in Africa differ sub-stantially from those in the countries of destination. This may lead to effects that are unknown in domestic systems. The main example we found is the process of rapid evacuation.

The neck of the hourglass model deserves special attention. In studies of domestic systems, it has long been recognized that this narrow neck represents a concentration of actors and stocks. It is here that economic power tends to be concentrated. Moreover, gov-ernments tend to intervene at this stage. But where is the neck located in trans-oceanic systems? In the oceanic section of the TOMC or in the ports on either side? We are used to the idea that effective power is located within national territories and that there is a power vacuum on the oceans. However, this idea is misleading when applied to the TOMCs. Table 15.1, which summarizes points from earlier chapters, suggests that there is some control over the product while it is in transit on the oceans, but that most actors seek control over the flow from a base on shore. The table may serve as a program for extending the theories of agricultural marketing.

With regard to public control at sea (where obstacles are very great), the table shows two attempts. While bulk buying by the British government (see Chapter 7) was a temporary measure taken

TABLE 15.1. Control Over the "Neck of the Hourglass" in Agricultural TOMCs

	Business Control	Public Control
On Shore	— trade associations — commodity exchanges — importer-merchants — international banks	— import and export regulations — taxes and subsidies
At Sea	— shipping companies — TNCs depending on special ships	— wartime bulk buying — international commodity agreements

during World War II, the postwar international commodity agreements (beginning with coffee in 1962) were designed as permanent schemes of control. Here, I want to stress that the commodity agreements were an extension of earlier private efforts, such as those of the trade associations. It is significant that the International Coffee Organization (ICO), for instance, decided to be guided in its policies by the price quotations on the commodity exchanges. Today these agreements have few supporters because the idea of regulating the prices in the world market (or in any other market) now seems futile or counterproductive. However, the other element in the design of the agreements—control over the middle section—should not be forgotten.[1] Such agreements may be revived in modified form in response to the globalization trend which puts a premium on administrative control of the oceans.

International Trade Theory

Exports constitute the common focus of international trade theory and the findings of this book. The classical theories of international trade contend that the country with the lowest production costs is the most efficient producer and will hold its own in international competition. In sophisticated studies, the internal marketing costs are added to the production costs to assess the competitiveness of, for instance, the coffee sector of a country (Daviron and Fousse, 1993). This model is upset by my findings about the *external marketing costs*. In Chapters 7 and 11, we found that some of these costs can be shifted from the country of origin to the country of destination. Moreover, in Chapter 11, we found that the amount involved is largely unpredictable because for many crops, the transaction costs have not been standardized. Hence, external marketing costs cannot be converted into a cost item in the theory of comparative costs and the analytical outcome of the international trade model is indeterminate.[2] The possibility of external marketing costs being shifted from one country to another is most interesting when a new TOMC is created. We found in Chapters 11 and 13 that the initiative often came from an importer outside Africa. Presumably, this was associated with a shift of external marketing costs from Africa to the countries of destination and a corresponding lowering of export prices.

In Chapter 14, we found that exporter preference for crops with a high degree of marketability tends to introduce a bias in the export basket. This is a significant microelement, which requires a modification or revision of international trade theory. Concretely, the circumstances which allow exporter preference to operate should be further studied.

While in existing theories, the FOB price for most export crops is seen as a variable beyond the control of the actors in the exporting country, we found that the exporters can influence it. They can raise it by building up a good reputation and by tactical speculation. However, we also found that exporters are prepared to accept a lower price if this reduces their stockholding burden. True, it is hard to measure these price-raising-and price-lowering effects because world prices, particularly those in the commodity exchanges, are permanently changing—an obstacle that has frustrated many researchers. Of course, the measurement problem does not relieve us from the task to take a new look at the FOB price.

Finally, the new findings highlight significant differences between the commodity exchanges and the theoretical world markets. How significant is the fact that the commodity exchanges are investors' markets as well as product markets? How significant is the fact that for most crops, it is impossible to establish a commodity exchange? Are there specific crop attributes to explain this difference? In short, the way in which the institutional world markets deviate from the theoretical ones should be further analyzed, even if it requires a substantial revision of international trade theory.

Development Economics

Once again, exports form the area of overlap. According to existing theories, a country may experience economic development in the field of exports as a result of (a) the importation of capital and skilled labor, the mobile factors of production; (b) the acquisition of new knowledge; and (c) government intervention. I have nothing to add to the first point point except that the scope for development along these lines is limited because suspicion of foreign enterprise is deep-seated in Tropical Africa (see below).

As to new knowledge, existing theories primarily see it as new and superior technology. My approach has emphasized another aspect: growing commercial insight. This type of new knowledge is expressed in the concepts of emancipation (Chapter 9), sophistication (Chapters 11 and 12), and countervailing power. While these concepts are primarily applied to the upper section of the TOMC, a wider perspective is necessary. The countervailing power of growers or middlemen may force an exporter to seek compensation farther down the channel and so raise the effective FOB price.

For several decades, government intervention was seen as the principal factor in development, but when this view was replaced by the views of neoclassical economics in the 1980s, the role of the African governments was drastically reduced. We will return to this point in the final section.

Economic History

Economic history recognizes the importance of the economic changes which occurred during the second half of the nineteenth century, as does this study. With these changes, the term "world economy" became meaningful for the first time. More helpful, I believe, and more precise, is the term "international economic order" introduced by Arthur Lewis (1978b). The term has the advantage of evoking the structural change rather than merely referring to geographical extension. Lewis carefully studied the period of 1870 through 1913 to see how and when this new order evolved (Lewis, 1978a). Like him, I am fascinated by the process of global economic integration. New technology was obviously important, for instance, in the area of ocean transport, where modern ships facilitated trade. Technology was important in the field of communications too; the submarine telegraph cables made a crucial difference.[3] Analytically, there is not much difference between technical and institutional innovations.[4] Yet few scholars of African history have stressed the new institutions to the same extent as I do.

Was the new international economic order characterized by a permanent division between agricultural and industrial countries? Have the former since been at a disadvantage in international trade because of an inherent inferiority of agriculture?[5] My findings, even if restricted to Africa, bear on these questions because I suggest that it was the new institutional world markets created in the nineteenth century which facilitated the export of agricultural (and other primary) products to a greater extent than that of industrial products. Indeed, my findings may be used to argue that it is more difficult today for a country to switch from primary to secondary exports than it would have been before these markets were created. My view on institutions helps to explain why there was so little change in the "international economic order" after the African countries gained independence.

I further hope that my findings will raise new questions to historians engaging in research on Africa's trans-oceanic trade in agricultural products. Until now, nearly all economic research was inspired by neoclassical economics or Marxist theories. As I see it, the framework offered by these theories precluded attention for these questions.

IMPLICATIONS FOR FAIR TRADE ORGANIZATIONS

It is helpful to begin with the concept of the *farmer's share*, a generally accepted way of measuring farmers' incomes. The farmer's share may be defined as the portion of the consumer's outlay that the grower receives on average, expressed as a percentage of this outlay.[6] Measuring the farmer's share can be easily incorporated into MCA; the calculation is easiest when the product sold by a farmer is identical with that bought by a consumer. Good examples are eggs and bananas.

The share of consumer outlay that does not go to the farmer is received by the *marketers*, a term covering traders, processors, transporters, distributors, etc. It has come as a surprise to many people that the farmer's share is low in relation to that of the marketers, even in a country such as the United States. For 74 domestic food items sold in the United States in 1987, Kohls and Uhl (1990, p. 193) found the farmer's share averaged 30 percent. Kohls and Uhl noted that it varies over time and among food items. Thus, for bread, the farmer's share was only 7 percent in 1987, but for eggs, it was 62 percent (ibid. p. 195). Kohls and Uhl identify several factors that lead to a low figure: (1) degree of processing, (2) perishability, (3) seasonality of the harvest, (4) high transportation costs, and (5) bulkiness. As a sixth factor, I suggest the velocity of the product flow. If this is low, the marketers' cost of finance and risk-bearing are high and the (complementary) farmer's share is low.

The same calculations can be made for products that move in a TOMC. There are no analytical problems (unless there is foreign exchange control), but the practical problems are considerable. Figures are therefore relatively rare. Clairmonte (1975, p. 138) puts the farmer's share for bananas in 1971 at 11.5 percent. The farmer's share for coffee was 27.6 percent in Colombia in 1979 and 8.9 percent in Côte d'Ivoire in the 1976-1979 period (Clairmonte and Cavanagh, 1988, p. 31).

A related measurement is that of comparing the price a farmer receives with the FOB price an exporter receives. While for grower-exporters these sums are identical, for small farmers (as studied here), there may be a substantial margin between the two amounts. Measuring this margin gained in popularity after the EMBs had been established, when it became possible to relate the widely publicized producer prices to the FOB prices in the foreign trade statistics. One of the first to make such calculations was Bauer (1954). More recently, Bates (1981) calculated such figures for eight crops in seven African countries.

Measuring the farmer's share has nearly always been associated with considerations of fairness, first in the West and later in the Third World. Does a farmer get a fair price for his/her product? Or, put differently, does a marketer obtain more than he/she deserves? In Africa, profarmer feelings and antitrader sentiments have frequently gone together. Allegations of unfairness were already being made in the decades before World War II when exporters and middlemen were accused of exploiting the farmers. Because hard figures were lacking, these allegations could not be substantiated.

After the war, the new EMBs were also accused of exploiting the farmers. Such charges could now indeed be substantiated by calculations of the type referred to above, using the FOB price rather than consumer outlay as a basis. The interesting point about these accusations is that the government was pilloried as being unfair. The EMBs were primarily discussed as fiscal instruments, so much so that their role as marketer was often passed over. Moreover, it was frequently forgotten that the precepts of fairness subscribed to by government differ from those a marketer might hold.

In the 1970s and 1980s, the issue of fairness surfaced in another area. Some consumers in the West came to the conclusion that they were paying too little for many products originating in the LDCs. To remedy this situation, they established Fair Trade organizations. The task of the Fair Trade organizations was (a) to study the producer's share in the LDCs and (b) to take measures to raise it.

The conclusion that the producer's share was too low was most easily drawn for a few agricultural products such as coffee, tea, and cocoa, because those products undergo little transformation between production and consumption. It is not surprising that these products loom large in the programs of Fair Trade organizations. (For rubber and cotton, it is much more difficult to calculate convincing figures.) The policy objectives of the Fair Trade organizations differ according to whether crops are grown by small farmers or on plantations. In the first instance, the aim is to raise the price received by the farmer (or by his cooperative). In the second it is to raise the wages of the plantation workers.

Since the international marketing channel, which is implicit in fair trade thinking, corresponds to the TOMC concept developed in this book, some comments, however brief, are in order. First, I endorse the basic idea that average incomes in the LDCs should be raised and that financial transfers from the rich countries are necessary for this purpose. Second, the fair trade idea that these transfers should be made within particular TOMCs is an arbitrary decision. I am prepared to endorse it if certain conditions are fulfilled. Thus, if coffee farmers or laborers on tea plantations deserve more financial help than the other inhabitants of the country, selective transfers are justified. Third, the policy of the Fair Trade organizations to buy straight from farmer cooperatives is only fair if the preconditions for the establishment of cooperatives are the same in the various countries of production. In Africa, this does not seem to be the case. These preconditions certainly deserve further research. Fourth, the tacit suggestion that trade, as presently organized, is unfair and has to be reformed betrays antitrader sentiments which, as stated, frequently accompany profarmer policies. I would like to challenge the Fair Trade organizations to reexamine the role of the traders (in the widest sense). Perhaps this role is more complex and justifies a higher marketers' share than the Fair Trade organizations think. This book demonstrates the complexity of trans-oceanic trade.

IMPLICATIONS FOR AFRICAN GOVERNMENTS

We must finally consider the relevance of the new findings for African governments. Here, six general, but brief remarks are necessary. First, the significance of the agricultural export sector varies greatly among African countries. Thus, countries with large oil revenues can afford to neglect export agriculture. Second, a government's attitude toward trans-oceanic crops is important. As explained in the first chapter, I assume that the governments favor these crops: governments are committed to export-led development and are aware of the high potential of trans-oceanic markets. Third, the governments' role has been greatly reduced since the early 1980s as part of various structural adjustment programs. Participation in the economy (including participation in the TOMCs for export crops) is now considered undesirable. However, a government remains responsible for its national economy in a limited way: it must study economic problems, advise private businessmen, facilitate development, and take a stand in controversial issues.

Fourth, the agricultural export sector is too large and unwieldy for it to be dealt with as a single entity. A government should therefore divide it into a number of subsectors. One way of dividing it follows from the division into four types of trans-oceanic crop introduced in Chapter 5. This leads to four subsectors which were compared in Chapter 14 in connection with export diversification.

Fifth, African governments sometimes have to clarify their stand toward foreign enterprise. In this connection, I must warn not only against antiforeign sentiments, but also against a common but mistaken perception of foreign influence. It is often limited to the notion of a foreign presence, that is, foreigners residing in the country and foreign enterprise operating there. Foreign participation is thus noticed and measured insofar as it is visible in a TOMC upper section. However, this perception is too narrow. For instance, a long-term purchasing contract, creating a vertical ocean-straddling marketing system, may give an importer in Europe extensive power in Africa. Moreover, it is inherent in the nature of a TOMC that the consumers are foreigners who live far away. Even so, the sector in Africa is dependent on them.

Sixth, African governments are occasionally asked to take sides in the debate on large- versus small-scale cultivation of trans-oceanic crops. This debate usually centers on feasibility studies which permit a comparison of the costs of the two forms of production and create the impression of objectivity. In practice, there are subjective factors as well: the desire to control production and the preference for modern production methods. Let us therefore recall the advantages of small-scale cultivation which we discovered because of the MCA approach: it spreads risks over the actors in the sector and makes the sector more resilient and viable in international trade.

Reorganizations in the Upper Sections

African governments are often confronted with demands for the reorganization of the TOMCs, demands which may come from actors within the channel or from outsiders. Recently and in keeping with structural adjustment policies, the focus has been on reorganizations which aim at privatization, competition, and the reduction of marketing costs. This is valid and useful, but incomplete. Governments should consider at least four other factors.

First, a government should support actors who are able to raise their selling prices. This holds not only for the farmers (in which case everybody agrees) but also for the exporters (as explained in Chapter 7) and even for the intermediaries because their strength prevents complacency among the exporters. One implication I want to single out: a government should (a) support strong exporters and (b) oppose reorganizations which weaken them.

Second, because of the high convergence factor, the power of the downstream actor tends to be great. Fear for this power is widespread among upstream actors and should be explicitly recognized, notably in discussions with foreign consultants who may not be aware of it. Competition is the standard instrument to keep this power in check. More promising, I feel, is the commercial emancipation of the upstream actors and the accompanying countervailing power. If such emancipation leads to the establishment of a cooperative, a government should welcome this, but the cooperative is no goal in itself. If it is not based on commercial emancipation, it may well be an empty shell.

Third, many actors in the channel are willing to make sacrifices to preserve the continuity of the channel and to build up a good reputation for themselves—two related objectives. Governments should welcome this, but should also realize that brushing up one's own reputation may be accompanied by attempts to discredit other actors, either at the same or at another level of the channel. This makes it difficult for a government to be impartial when examining reorganization proposals put forth by actors.

Fourth, reorganizations may be counterproductive. They may retard or upset the slow processes of emancipation and reputation building and may partly destroy what has been achieved. The negative effects may be strongest for the upstream actors. Indeed, many reorganizations, whether introduced by the ministry or by private enterprise (with the consent of the ministry), may cause serious frustrations to the farmers. To avoid misunderstanding, I am not defending the status quo under all circumstances, but the initiatives should come from upstream and as a result of genuine emancipation. In this connection, it is important to initiate research on the modification or replacement of traditional commercial practices. Such research would also offer a good starting point for studying the commercial emancipation of the farmers.

CONCLUDING REMARKS

Because of my affinity with the policymakers, I have felt a sense of urgency in writing this book. I want to help prevent unrealistic reorganization plans from being launched. I also hope that African governments will avoid wasting money and manpower in misguided directions.

I also want to make it clear that I am not entrapped in the gloomy determinism of many global theories about Africa and the world economy. The economic history of this continent shows that there have been times of surprising growth and effective innovation. It is my hope that this book will help to revive academic and political interest in economic development and to combat current pessimism.

Glossary

external marketing costs: Costs related to marketing operations in the middle section (and possibly the lower section) of a TOMC.

farmer's share: The portion of the consumer's outlay on a product that the farmer (grower) receives as a percentage of this outlay.

marketer: A collective term for all nonfarmers in the marketing channel, such as traders, processors, transporters, and distributors.

Notes

Chapter 1

1. Normally country of shipment and country of origin may be equated, as well as country of consumption and country of destination, but landlocked countries complicate the analysis. It is generally agreed that these countries are at a disadvantage because they are restricted in their trading partners and routes.

2. There were few answers in the secondary sources I consulted. It is, however, possible that the primary sources contain far more answers. Such a discrepancy could be accounted for by the fact that the theories that guided economists and economic historians in their research have been one-sided.

Chapter 2

1. French scholars use *filière*, a term which corresponds to marketing channel.

2. Galbraith (1957), who introduced the concept of countervailing power, was the first to emphasize relationships across the market.

3. Hancock stresses the distinction between expansion overland, as in the United States, and expansion overseas. See particularly Hancock (1943).

4. This area should not be confused with *Afrique Centrale*, which is much farther to the North and is part of Atlantic Africa.

5. For the relationship of exports and imports in a peasant economy (value-wise and season-wise), see Van der Laan (1991).

6. In Africa, where the term "intermediary" has a narrow customary usage, it would be confusing to follow the MCA practice of using the term for each channel member.

7. I had this metaphor in mind when I constructed Table 2.1. Whereas the transactions of Group 1 represent the moments of handing over the baton, the intervals of Group 2 correspond to the intervals when the athlete is running.

8. By going back in history, it is usually possible to say whether the initiative came from the countries of origin or from elsewhere.

9. Protection based on import tariffs could, however, not be used in the Congo Basin, a large area in Central and East Africa, because of a multilateral treaty of 1885.

10. Under competition, there are several exporters, each of which has to build up his own channel system.

11. The image of the river drainage system may be further simplified to the model of an inverted pyramid. The pyramid's height is determined by the number

of successive actors, and its width at the base, by the convergence factor. This model is helpful in the analysis of Chapters 6, 8, and 9.

Chapter 3

1. Often the terms "shipping line" and "shipping company" are used interchangeably, but a line does not have to register as a company.

2. Governments encouraged modernization by awarding mail contracts to the most dependable line.

3. Such warehouses have been rare in the ports of shipment (see Chapter 10).

4. The French, more appropriately, speak of a *ristourne de fidélité*.

5. As the lines broke with the old policy of waiting until the ship was full, some ships had to sail partly empty, i.e., with a low load factor.

6. In the literature on Africa, many complaints about delays in shipment are recorded. Most of them concern delays caused by port congestion. If the shipping lines had caused similar delays by not making sufficient shipping capacity available, the exporters would have complained no less vocally, I feel. However, I found little evidence of this. This suggests that the shipping lines serving Africa were quick in adjusting capacity.

7. These are used mainly for iron ore, which was not exported from Africa before 1930. The export of nonferrous ores such as tin and manganese began much earlier. Those ores were carried in general cargo ships.

8. Public investment has been substantial in many countries because the authorities realized that such warehouses made the port attractive to the shipping lines.

9. The most recent closure lasted from 1967 to 1975. This greatly lengthened the distance from the East African ports to Europe. For shipments to and from the United States, the closure made little difference, because the American lines had a combined route for East and South Africa.

10. The Union Castle Line operating between Great Britain and South Africa may have served as an example. Their ships had always made straight sailings.

11. In the 1920s, the Conference was joined by a Dutch line and was renamed West African Lines Conference. In the 1950s two Scandinavian lines joined, and in the 1960s two African lines (from Ghana and Nigeria) joined.

12. On this route, specialized African workers, most from the Bassa and Kru tribes, were employed. They boarded the southbound ships in Freetown or Monrovia, did the stowing of cargo there and at subsequent West African ports, and disembarked on the northbound trip.

13. Insufficient tanker capacity for bulk oil transport was another grievance of UAC. The fact that no rebate was paid on cocoa, cotton, rubber, and timber was probably also influential.

14. Leubuscher, 1963, p. 50. She cites UAC's own words, . . . "rather than submit to the domination and dictation of the shipping monopoly" from a cablegram of December 1929.

15. Around 1935, UAC decided to increase its own fleet. By 1939, it owned fourteen ships and two tankers (Fieldhouse, 1994, pp. 179–181). However, char-

ter ships continued to be used for temporary excess cargo. In 1949, UAC created a separate company, the Palm Line, and transferred all of its ships to it.

16. This had been conspicuous in the late nineteenth century, when small African and European traders profited from the opportunities created by the lines (Hopkins, 1973, p. 151).

17. Northbound cargoes tended to exceed southbound ones in West Africa—a normal feature of an agricultural region. Exporters who operated their own ships could give priority to high-value crops such as cocoa and postpone the shipment of other crops, such as palm kernels, until the slack months.

18. John Holt & Co., a trading company from Liverpool operating in Nigeria, had benefited from this possibility since the 1860s.

19. Leubuscher, 1963, p. 48.

20. In 1940, an American line began to use tankers to transport latex from the Firestone plantations in Liberia to the United States (Albion, 1959, p. 205). This replaced the export of latex in drums and, to a lesser extent, the export of dried sheet rubber.

Chapter 4

1. To meet the wishes of the customs officials, trans-oceanic CIF contracts are normally drafted so as to make it easy to calculate the FOB price.

2. I was first alerted to their significance by the cable addresses on the letterheads of the early exporters and by their fascination with (unbreakable) codes.

3. This premium helped compensate for the discount on forward sales previously mentioned.

4. The TOMC, as used in this book, presupposes the telegraph system. Historians may argue that I ignore the earlier TOMCs. This is true, but these early TOMCs were primitive and operated in a very different way.

5. Belgium was not a colonial power until 1909. The Congo Free State of the period before 1909 relied on an overland cable from Luanda.

6. See Lesage (1915) for more information on the German cables. During World War I, the German cable to Monrovia was cut by the Allies (Buell, Vol. II, 1928, p. 767).

7. The new Bank of British West Africa opened its first branch in Africa (in Lagos) in 1894. The National Bank of India opened a branch in Mombasa in 1896 and the Banque de l'Afrique Occidentale started its operation in French West Africa in 1901 (Hopkins, 1976).

Chapter 5

1. There are various definitions for "transaction costs." I prefer the short one given by Kohls and Uhl (1990, p. 536): the costs incurred by buyer-seller search, negotiation, and contract enforcement activities.

2. The broker is not merely an intermediary. He is also responsible when his principal defaults.

3. Because of the different time zones, the opening hours of the London and the New York Exchanges hardly overlap, which reduces duplication.

4. Terminal markets never provide long-run price forecasts. They have elaborate rules about clearing, margin calls, halts to trading, etc.

5. Hedging techniques are described in many textbooks.

6. Traders speak of "stocks on hand" and "stocks in sight." The latter correspond to the quantities given in the crop forecasts, e.g., those prepared by the U.S. Department of Agriculture.

7. The regulations governing these markets are drafted in such a way as to condense all variability into the price. Such condensing of market intelligence into one figure is possible for only a few types of crop.

8. Both World Wars brought the cessation of trade and the closure of the exchanges. Even local wars cause abnormal price movements.

9. Businesspeople may prefer private deals when, because of the integrity of their trading partner(s), problems of international contract enforcement are negligible.

10. Exchanges for new products are regularly launched; many fail because too few traders switch from their customary practice of making private deals to making public contracts on the exchanges.

11. Many terms (class, type, and quality) are used, often without an indication of the precise meaning. I prefer the term "class," which I contrast with the term "grade."

12. Presumably, some exporters sold their product on description and thus in undifferentiated form. This meant less revenue, but an earlier sale.

13. There are two well-known exceptions in Africa. First, Firestone, the American car tire manufacturer, established a large rubber plantation in Liberia in the 1920s. We should remember that Firestone was afraid that the Stevenson Scheme would curtail its supplies from Asia. Second, Lever Brothers established oil palm plantations in the Belgian Congo before World War I. At that time, the exchanges for oilseeds were still small. Moreover, William H. Lever was known for his preoccupation with secure supplies.

14. If the exporter and grower are separate actors, the question must be raised whether high marketability also benefits the grower (see Chapter 7).

Chapter 6

1. A favorite expression of those holding this interpretation was "enlarging the exchange economy."

2. The sugar and oil palm plantations which were established in the same period produced for the domestic market and are therefore not relevant for this book.

3. In the nineteenth century, this view led to the use of Asian indentured labor on the sugar plantations on Mauritius and in Natal. The use of such labor was seriously considered in British West Africa (Phillips, 1989, pp. 37–39).

4. The literature abounds with accusations of false weights and scales that were tampered with. We must realize, however, that even if all weighing was correct, the foreign contraption would still have caused suspicion among the growers!

5. Van der Laan, 1987b, p. 16. Buying at the farm has been exceptional in the upper sections of the TOMCs. As a result, the farm gate price is not a very useful concept in Africa.

6. In Figure 6.b1, the convergence ratio shown is three growers to one exporter. In practice, the ratio has been much higher: about fifty to one. The higher the ratio, the more cumbersome negotiating was.

7. Buyers tended to cheat by using false weights and growers by adulterating the produce.

8. My explanation for the practice is not the only one. A popular explanation has been that the African farmers were unwilling to wait for the money they were entitled to. (This was attributed to an extreme lack of cash in the rural areas and/or to the farmers' improvidence.) Perhaps the anthropological literature contains fragmentary information about farmer's relationships with traders. When was the relationship considered personal and when was it considered anonymous? Or, in terms of the New Institutional Economics, expectations become more important in economic behavior, as relations become more personal. Presumably, this observation can be applied to internal marketing in Africa.

9. Sorting by the grower is important. It must be done when the grower combines the elementary botanical units such as cocoa beans, coffee cherries, cotton bolls, or unshelled groundnuts into parcels. In general, it pays for the grower to ensure that his parcels are as homogeneous as possible.

10. I distinguish outgrowers from contract farmers; for the latter, see Chapter 8.

11. To simplify the argument, we assume that all of the intermediaries are private businesspeople, but in some areas, cooperatives have functioned as intermediaries. For cooperatives, see Chapter 8.

12. The intermediary simply had to refuse substandard produce or "trash" (Pedler, 1955, p. 83).

13. The number of growers per intermediary was generally lower than that per exporter in Pattern 2.

14. In the years of anonymous direct trade, rural indebtedness could not have arisen, but in the 1930s, it was substantial in many areas and a reason for concern for the colonial authorities.

15. Cooperatives were often recommended and established to help the small growers, but in practice, the influential growers retained a favorable position.

16. A reliable but bizarre solution would be to mark the botanical units from 100 parcels at various PMPs, to trace them at the port of shipment, and then to calculate the average time they had taken to get there.

17. Several other metaphors have been used to illustrate the pattern of convergence. Pedler (1955, p. 151) speaks of a fan; Suret-Canale (1964, p. 251) speaks of a lung; and Johnson (1970) mentions a tree and a dendritic system.

18. In some countries, the construction of new roads, notably in the 1950s, altered the transport pattern. As a result, the old and the new drainage patterns may not be identical. This is a special challenge for geographical analysis.

19. This metaphor introduces a strong deterministic element into the analysis. This is counteracted by the metaphor of the relay run (also introduced in Chapter 2) which stresses human effort, skills, and freedom of choice.

20. Chad and the Central African Republic now export their cotton via Douala in Cameroon. During most of the colonial period, they shipped via Matadi in the Belgian Congo and, after 1934, Pointe Noire in the French Congo.

21. Occasionally, geographic information is essential for understanding channel systems. I discovered this in Sierra Leone, a country which had a second port, Bonthe, until about 1960. As long as the hinterlands of Freetown and Bonthe were separate, it was possible for an exporter in Freetown to control his own channel system in Freetown's hinterland. However, an exporter risked losing control when new roads were built, as happened in the 1930s. These roads enabled intermediaries to shunt crops to the hinterland of Bonthe (Van der Laan, 1975, p. 133).

Chapter 7

1. A shorter version of this chapter was presented at the Forty-Seventh Seminar of the European Association of Agricultural Economists in Wageningen, Netherlands, March 1996.

2. Cablegrams were important in the past, both for negotiation and confirmation.

3. This presupposes reliable international telecommunications.

4. A sound reputation raises his effective price compared to those obtained by rivals with a lesser reputation. The margin, although small, may be significant in a sector where the trading margins are narrow.

5. Since physical stocks can never be negative, physical and economic stocks have to be carefully distinguished.

6. The analysis needs to be modified for rubber and oil palm, which are harvested during most of the year, and for cocoa, which has a small so-called "mid-crop" during the off-season in a few regions.

7. Processing, for instance, cotton ginning, may reduce the gross volume, but the effective volume is indeed constant. Value is added during processing, but this is not taken into account in the diagram, which measures volume.

8. This shift of price risk assumes that sale and shipment coincide—an assumption which will be abandoned later in this chapter. Moreover, a fixed-price contract is assumed. The price-to-be-fixed contracts came into use in the 1970s only.

9. These sales are not indicated in the diagram for the practical reason that an exporter can follow alternative scenarios: he can either keep his stocks in the port of shipment until a buyer is found, or ship them unsold, as in the days of consignment sales.

10. These points are marked on his calendar, so to speak. Importers monitor information on transport problems because they may shift point F beyond its normal position.

11. The terms "nearby" and "distant" are primarily used in the terminal markets. When I extend them to characterize shipment, the fit is not perfect, which is why I use the term "roughly" as a qualifier.

12. A contract that specifies shipment in about six months' time may serve as an example.

13. Later, at the moment of shipment, the exporter may regret his speculation. This may happen with all forward sales, as we saw above.

14. Contango premiums are less attractive for importers because importers have to pay storage costs for the whole period. The exporters' storage costs are lower because they take delivery at a later date, closer to the moment of shipment.

15. This conviction is especially justified when a country is a large supplier and when the commodity has a global seasonal pattern, such as cocoa, which ripens in many countries at about the same time. There is a parallel with domestic markets where concentrated postharvest sales are considered less than optimal for the farmers. However, we must remember that farmers are advised to sell *after* the harvest, while exporters (and exporting countries) are advised to sell *before* the harvest. The practice of CIF trading is responsible for this difference.

16. The general manager had a ledger into which he entered purchases and sales every day. If the balance was zero at the end of the day, he had squared his book.

17. To scholars of West African history, it must seem incredible that companies which were so huge in the African context seemed to fearful and timid, not to say cowardly. However, because of world price volatility, stock depreciation in one season could easily wipe out the trading profits of several years.

18. The design of the pool suggests that the exporters were not worried about delays due to logistic problems between the IMPs and the port. This is a tacit but significant comment on the economic conditions in West Africa.

19. In the Nowell Report, only the disadvantages of being such a megaexporter are mentioned: UAC could not take refuge in a risk-avoiding policy like its rivals, but had to face the price risk of a short position at first, followed by that of a long position later. Of course, such a dominant position also conferred benefits!

20. In a document of 1931, UAC distinguished the market profit (which corresponds to my moment-making profit) from the straight-through profit (Fieldhouse, 1994, pp. 113-114).

21. When shipments stagnated in 1938 as a result of an unprecedented cocoa hold-up in the Gold Coast (see Chapter 9), UAC almost had to buy cocoa elsewhere to avoid defaulting on some of its forward contracts (Fieldhouse, 1994, p. 168).

22. Such stabilization was limited in scope: the government was not supposed to subsidize the sector if it became uneconomic.

23. I refer to the *Caisses de Soutien* for cotton in Chad and Ubangi-Shari. They made support payments to the licensed ginner-exporters during the years 1931 through 1936 (Cabot, 1965, p. 200). The "search for security" manifested itself earlier in the French colonies than in the British colonies (Hopkins, 1973, p. 264).

24. A lack of symmetry should be pointed out. A public body which takes on the short-run price risk automatically takes on the long-run risk as well. The

opposite is not true. The early *Caisses de Soutien* are a good example. They bore part of the long-run risk, but avoided the short-run risk.

25. The grower-exporters normally remained outside the EMBs (but two cocoa plantations in the Southern Cameroons came under the Nigerian EMB). There are two explanations for this situation, both of which plausible: (a) the colonial authorities were unwilling to extend their protection to the mainly foreign exporter-growers; (b) the exporter-growers wanted to retain their freedom.

26. Bauer contrasted the prewar situation of free trade with the postwar monopoly situation, but, as will be shown in Chapter 9, the two situations differed much less than Bauer thought.

27. Leubuscher (1956) provides useful information on the implications of bulk buying for the EMBs.

28. This could also be a disadvantage to the farmers. Cautious EMB managers may have occasionally delayed the fixing of the producer price if the delay helped to reduce the period of exposure to the price risk. In my opinion, the influence of risk-averse EMB managers has been underrated, except by Leubuscher (1956, p. 95).

29. Not only could the acreage sown not be determined, but several annual crops such as groundnuts and cotton were also grown in areas with highly variable rainfall.

30. In most countries, the statutory monopoly was also a *de facto* one, but in some countries, e.g., Senegal and Ghana, part of the crop was often smuggled to neighbor countries.

31. As the *Caisse* did not buy or hold stocks, it entrusted the actual preparation of the contracted consignments to licensed exporters against a fee.

32. A common formula fixes the effective price equal to the spot price quotations of the same day, minus a margin or differential agreed upon in the contract. Obviously, such a contract turns the exporter into a moment maker.

33. However, Eric Tansley's loyalty was to the EMBs and the farmers and not to UAC and the other private exporters (Fieldhouse, 1994, p. 275). See also "Cocoa Man," *West Africa*, 1961, p. 653.

34. Twenty years later, Eric Tansley was still remembered for this achievement in Dutch cocoa circles (Interview of the author with F. Fopma, former director of CONTINAF, December 1982).

35. In the early 1960s, the new telex facilities in West Africa brought an organizational change. Formerly, the selling decisions had been in the hands of EMB employees in offices close to the exchanges in London. Now, the selling decisions were made at the EMB headquarters in Africa. This suggested, powerfully but erroneously, that the EMB had a much higher degree of market control.

36. This term also covers the public exporters in the francophone countries such as the *Offices de Commercialisation*.

37. This was not part of a donor-proposed structural adjustment program, but an internal, Nigerian policy.

38. There was a thinly disguised hostility toward those who had run the EMBs. It was alleged that they had abused their position of trust and had lined their pockets

without pursuing the interests of the enterprise or the country. The fact that the export transactions made by the EMBs were private deals fostered suspicion. Thus, the call for transparency was a powerful element in the privatization drive. However, these criticisms ignore the advantages of private deals and the fact that this kind of abuse may just as well occur in private export organizations.

39. A detailed comparison of the benefits of the old forward sales contracts and the new "put options" seems desirable.

40. If farmers are paid in two installments (see Chapter 8) and thus bear price risk after delivery, the exporters provide some cover against this risk under certain arrangements (Claessens and Varangis, 1993).

Chapter 8

1. I have chosen this term because it runs parallel with the political term "indirect rule."

2. Most cotton seed has been distributed free of charge. The system of distribution is significant. The ginner can choose for direct distribution to each farmer individually or for indirect distribution through the chief or the cooperative.

3. Storage costs may be further reduced by working day and night, as was done in the ginneries in Uganda in the 1930s (Thomas and Scott, 1935, p. 139).

4. Van der Laan (1984) briefly reviews the operations of the Nieuwe Afrikaansche Handels-Vennootschap (NAHV) in the Belgian Congo and Ubangi-Shari between 1918 and 1955. In the history of its export marketing, three stages may be identified. At first, the NAHV made consignment sales. Next the company made forward sales and protected itself against the price risk by selling futures contracts on the New York Cotton Exchange. In the third stage, the NAHV also protected itself against the currency risk.

5. For cotton in Mozambique, see Isaacman (1996). For cotton in Sub-Saharan Africa in general, see Isaacman and Roberts (1995).

6. The account given here rests on the (1959) authoritative book by Gaitskell.

7. See the prewar account by Thomas and Scott (1935, pp. 133–140) and the postwar review by Ehrlich (1965).

8. Only part of the cotton is exported; the remainder is sold to local textile factories.

9. I am indebted to Frank Bessem for the information in this section. He did field work in Northern Cameroon in the early 1990s. His PhD thesis will deal with the SODECOTON and the *Associations Villageoises des Producteurs* and the *Associations Villageoises Autonomes*.

10. Studies on cooperatives tend to focus on what happens at the PMPs. In general such studies ignore the problems downstream (at the IMPs).

11. The maximum reallocation of functions turns an exporter into an agent of the grower, who now bears nearly all risks and financial costs, but in return, receives the full value of the product.

12. In the Gezira Scheme, it did not matter whether the growers were sophisticated because the government of the Anglo-Egyptian Sudan monitored the performance of the SPS. (See Chapter 12 for differences of opinion between the two).

Chapter 9

1. Cf. Hancock (1942); Bauer (1954); Amin (1971, 1972) on the *économie de traite*; Hopkins (1973); Hart (1982); Munro (1984); and Austen (1987).

2. However, nationality was not decisive for a company's sphere of activities. Some British companies, for example, operated in French colonies and vice-versa. The German companies which were important before World War, I disappeared afterward.

3. We do not mention some older companies such as AETC and the Niger Company whose names disappeared when UAC was formed (Pedler, 1974).

4. The practice of spending at once had its origin in the days of barter.

5. Several companies had painfully learned that an overestimate was undesirable because the unsold merchandise would lie idle up-country until the next crop was harvested. There was a similar problem for the country as a whole. It was bad business for the companies to order and import too much (Van der Laan, 1991).

6. In general, the river stations were older. It is significant that the French term *escale*, originally used for river stations, was later applied to all buying stations.

7. Telegraph cables were laid by the railway organization to ensure safe and smooth operations of the trains. It meant additional income to open these lines to the public.

8. The river-based companies in Nigeria and Senegal declined in relation to the rail-based ones (see Chapter 10).

9. The trading companies felt they could rest on their laurels. However, I also mention a practical consideration. Having road stations entailed the establishment and operation of a fleet of lorries. From some interviews with traders in Sierra Leone, I got the impression that the companies were reluctant to operate lorries. Experienced drivers were scarce and difficult to supervise. Third party insurance was not available, and road accidents harmed a company's reputation.

10. We are reminded of Hancock's (1942) "frontier" concept. When Hancock spoke of the [European] traders' frontier in West Africa, he meant the companies frontier at the IMPs. He had no eye for the middlemen's frontier at the PMPs.

11. This was derived from the word "factory," originally used for trading posts. Other names such as "agent" were also used.

12. Economists have preferred to write about the agreements to restrict competition, the "pools," because those agreements fit into the economists' theoretical concepts. However, the tying arrangement is best explained in MCA terms.

13. Apart from credit, there were other privileges such as easy hire purchase terms for lorries or allocations of special textile prints.

14. Company credit enabled some middlemen to sell merchandise on credit to the farmers. This extension made the credit chain vulnerable, as became clear in

the crisis years of the 1930s. It was a replica of the tying system, bringing the farmers into a dependent position—an issue normally studied as part of the phenomenon of rural indebtedness.

15. Those that were so-called "strangers" (because they were Africans from another territory), such as the Lagosians in the Gold Coast, attracted more criticism.

16. For a recent evaluation, see Fieldhouse (1994).

17. There is a partial parallel with the exporter's moment making described in Chapter 7.

18. Most authors have highlighted the other aspects of the agreement, such as measures to restrict competition.

19. The innovation is here considered in the context of the relationship between exporters and middlemen. However, we should briefly look at the consequences which it would have had for the farmers. World price changes, even day-to-day changes, would be transmitted to the PMPs and the price risk would henceforth be borne by the farmers. The exporters would then lose their status as independent traders and become a kind of marketing agent for the farmers. Sooner or later they (and the middlemen) would give up their trading profits and be content with a standard commission.

20. For a similar view on the situation in Nigeria, see Deutsch (1990).

21. I learned a great deal about the EMB-LBA relationship when I interviewed Lebanese LBAs and subagents in Sierra Leone around 1970 (Van der Laan, 1975).

22. There had been a similar confusion about the legal status of the middlemen during the cocoa hold-up of 1937–38, as we just saw. The price stabilization introduced by the EMBs ended the opportunity for intermediaries to earn appreciation profits.

23. For the process of withdrawal and the reasons behind it, see Van der Laan (1983b), Deutsch (1990), and Fieldhouse (1994). Withdrawal from the produce sector entailed withdrawal from many locations up-country. This was favorable for the middlemen of the interwar period. Some of them could now graduate to the next downstream position in the TOMC and operate at an IMP. This economic advance, which I roughly date at 1960, was limited because they still had no experience of exporting and shipping.

24. However, access was not automatic. There have been a few instances of commercial banks refusing to lend more money to an EMB. This occurred in Sierra Leone in 1967 and resulted in a serious liquidity crisis which paralyzed trade. The actual burden was borne by the farmers who had to wait very long for their money.

25. We focus on the instrument and not on the EMBs, some of whom were weak, notably in accounting. At the end of the buying season, the accounts of the LBAs had to be brought up to date to see whether they still owed money to the EMB.

26. I noticed these problems in two provinces of Cameroon in 1987 (Van der Laan, 1988).

27. Not only was the carrot of high prices for good quality used, but also the stick of discouraging prices for poor quality.

28. In retrospect, this may offer an explanation for the reluctance of the prewar trading companies to use more than one grade in their buying operations.

29. I saw complaints with regard to Cameroon and Côte d'Ivoire. If the government stations produce inspectors up-country, it can easily collect information on the shares of Grade I and Grade II at the PMPs. It can later compare these shares with the corresponding national shares at the port of shipment. A considerable discrepancy suggests the practice of *mélange* (mixing). At the same time, we should note that the incentive for the licensed exporter to have good export quality, well above minimum quality, was retained under the *Caisse* system. This militated against mixing.

30. This malversation can only occur when producer prices are fixed. Since such prices were not introduced in francophone Africa until the mid-1960s, cheating with *lettres de voiture* (way bills) began only then.

31. Some EMBs were motivated by hostility towards the LBAs, not unlike the companies' hostility toward the middlemen of the interwar period.

32. There is an urgent need for a systematic examination of noncompliance, where noncompliance is defined as failure to comply with the instructions and expectations of the channel leader. Obviously, noncompliance occurs both inside and outside his organization.

Chapter 10

1. The groundnut sector in northern Nigeria illustrated this relationship well. Many people have admired the groundnut pyramids in Kano. This method of storage was developed as early as 1912 (Pedler, 1974, p. 169). It is significant that Abbott and Creupelandt (1966) used a photograph of the pyramids as the cover illustration for their book on marketing boards. It is nevertheless true that the pyramids represented a form of unwanted intermezzo storage due to the limited capacity of the Nigerian railways. Before around 1960, it normally took four months to evacuate the groundnut harvest from northern Nigeria.

2. A technical complication is that banks may not charge a flat rate throughout the year. Banks must earn their profits in the months of active agricultural trade. Furthermore, banks may impose penalties for late repayment (e.g., after six months).

3. However, newcomers find it hard to obtain bank credit—a point often noted in the 1920s and 1930s and now in countries where the marketing boards have been abolished.

4. The banks are even less reluctant to extend credit when an exporter is a foreign company and has assets outside Africa which may serve as additional collateral, as was the case during the colonial period.

5. For a banker, the amount of the export contract is a firm figure, in contrast to an estimate made by the exporter and based on the last commodity price quotations.

6. In the 1960s, the number of towns served by insurance companies, banks, etc., increased considerably.

7. Special ventilation techniques are required which influence the design of the produce stores.

8. The weight loss of palm kernels is discussed by Bauer (1954, p. 243).

9. Marine insurance does not cover quality deterioration, but thorough disinfection and fumigation in the port of shipment greatly reduce the risk. Historical

research on these activities is necessary. Nitrogen gas, which is an effective insecticide, but leaves no traces on the product, was introduced in the 1980s.

10. I heard about an interesting case during my research in Cameroon in 1987. Two years before, there had been a bumper cocoa crop. At the EMB's depot in Kumba, there were long queues of lorries because the EMB staff was unable to cope with all the cocoa that was coming in. The EMB took the complaints of the intermediaries seriously and reorganized produce receiving at this depot to prevent such delays the following year.

11. After having minimized their physical stocks, the exporters further reduce their burden by speeding up the legal title flow.

12. Miller Bros. moved in 1903 and the African Association in 1907 (Newbury, 1978, p. 559). The Niger Company moved from Burutu in the Delta to Lagos soon after 1920.

13. For details, see Van der Laan (1983a, pp. 560–562).

14. In Chapter 8, we noticed a strong tendency in the cotton sector of colonial Uganda to increase ginning capacity. I suggest here that the goal of rapid evacuation was largely responsible for this tendency.

15. The time perspectives of Chapters 7 and 10 differ somewhat. In Chapter 7, purchases at the PMPs were treated as taking place all on one day. These same purchases are now shown as spread out over two months. In Chapter 7, a microperspective was used. In this chapter, a macroperspective is required.

16. We must distinguish between weakness and invisibility. The process of rapid evacuation is hardly visible when production and shipment of a crop are spread over the year. Rubber is a good example.

17. The buying season may be defined as the interval between the first and last purchases made at the PMPs. It corresponds closely to the official buying seasons of the EMBs, but these continue until buying at the IMPs has been completed.

18. Measuring the seasons on the time axis is a modest objective. Plotting a curve which represents the reality corresponding to the triangle AGF of the model is more difficult. A proxy for this curve is the national seasonal finance requirements curve which bankers in many countries are able to plot.

19. In the francophone countries, data on the *masse monétaire* (volume of money in circulation) and its seasonal fluctuations have long been published (Durand, 1957). See also the statistics of the Banque Centrale des Etats de l'Afrique de l'Ouest.

20. This is true for only a few commodities such as cocoa, most of which is harvested at about the same time. Actually, world markets have a better chance to develop when the harvesting seasons fail to coincide. For instance, this happened toward the end of the nineteenth century when supplies in the temperate zones of the Northern Hemisphere began to be supplemented by those from the Southern Hemisphere. One of the advantages of such a world market is that it allows importers to switch from countries that are late harvesters to those that harvest earlier.

21. It is useful to consider the alternative situation: stocks remain at origin until the manufacturers need them. This situation is rare, but may occur if the commodity is grown in a rich country.

22. However, as we saw in Chapter 3, exporters may use their influence with the shipping lines to reduce the number of intermediate ports of call.

23. I expect these stocks to have a seasonal cycle, but this may be hard to document because of our statistical ignorance with regard to the trans-oceanic section.

24. It is assumed that most of the speculators live in these countries, but there is no proof. It is also confusing that many speculators often switch their attention from one marketing channel to another.

25. But the countries of origin automatically reap windfall profits when the price rise occurs before the stocks are sold!

26. In fairness, we should add EMBs they were not merely exporters. They had a general responsibility for the whole sector, as is partly explained in Chapter 7. This required a broader view which could easily distract from some of the commercial objectives.

27. Structural adjustment has often been advocated because of the discipline of the market which it brings. In the case of the public exporters, the discipline of the lender was more urgently needed.

28. There is a close link between physical performance and managerial competence, as discussed in Chapter 7. Both are essential for tactical speculation.

29. In the 1970s, when world prices were high, some landlocked countries exported their commodities by air. Thus, Rwanda and Uganda had some of their coffee flown out to Europe. Although this was only a temporary solution, it indicates how strong the goal of rapid evacuation is.

30. If the EMBs had been more aware of the stockholding burden borne by the licensed buyers, they would have conceivably hesitated before implementing plans to remove them from the channel. As it was, by going ahead with vertical integration (see chapter 9), the EMBs unwittingly assumed more of the burden while continuing to lack the drive to reduce it.

31. The companies involved are SOCOPAO and SOAEM. The facilities for such *nantissement* (borrowing against warehouse receipts) are mentioned in several French publications. The PTA (Preferential Trade Area) Bank fosters the establishment of such warehouses in Eastern and Southern Africa.

32. In countries where private exporters bore part of the stockholding burden, unscrupulous exporters were tempted to smuggle their unwanted stocks to a neighboring country. In the coffee sector, this was known as the problem of "tourist coffee."

33. This was already understood and studied around 1950 (Leubuscher, 1951).

34. Another problem is this: the processed product is less marketable than the raw material. Hence, greater marketing efforts are required and marketing costs are higher (see Chapter 14).

Chapter 11

1. Until 1930 there was a well-organized market in London with regular public sales (Stahl, 1951). Afterward, London remained important because of its concentration of influential merchants.

2. Sisal production in Tanzania declined greatly after the country's independence.

3. I was told that demand from the German steel industry rapidly increased because piassava fiber was found to be superior to other fibers in scraping hot steel. Apparently, its ash caused less contamination of the steel. There were two grades: Sherbro and Sulima. These geographical labels were not very accurate and may explain the lament of a retired UAC director (expressed in a conversation with me) that no product had given European businessmen in Sierra Leone more grey hairs than piassava.

4. Put simply, it takes a greater effort to sell a ton of, for instance, ginger than a ton of cocoa. Expansion of the marketing department is required if the board also wants to be in touch with a large group of potential clients so as to prevent large unsold stocks of a minor crop.

5. They would also recommend the establishment of trade associations and more global information. Valuable work on these issues is done by the International Trade Centre in Geneva.

6. Further research is necessary about the channel leaders. Did they operate from Zanzibar or from elsewhere?

7. The situation was similar to that prevailing in Sierra Leone's kola nut trade in the 1920s. AETC, the largest European exporter in Sierra Leone at the time, made it clear that it did not trade in kola nuts, a product which was shipped almost exclusively to African countries (Goddard, 1925, p. 193).

Chapter 12

1. When prices are high, growers have little incentive to differentiate their product. We should further not forget that doubts about the perfectly competitive market were strong at the time and inspired the theories of monopolistic and/or imperfect competition (Chamberlin, 1933; Robinson, 1933).

2. Many accounts of the tobacco auctions at Salisbury/Harare stress the folkloristic rituals.

3. Part of the reluctance was due to the failure of the syndicate's own auction attempt in Barakat in 1926.

4. Setting reserve prices is a delicate operation. Reluctance to sell through auctions may express itself in high reserve prices.

5. The local auctions permitted Sudan cotton to be directly sent in three different directions—west, east, and north—an improvement over the earlier system of sending the cotton to Liverpool.

6. Optional auctions were older, having been set up in Salisbury in 1910. An auction is optional, if growers are not compelled by government to sell their crops there.

7. Those African farmers belonged to the Chagga tribe and were grouped in the famous Kilimanjaro Native Co-operative Union (KNCU).

8. The KPCU was one of the few cooperatives in Africa set up by Europeans.

9. Even if the promotion of tobacco consumption is regarded as undesirable in many countries of destination, the positive effect on incomes in the countries of production should not be overlooked.

10. This is true for the tea trade in general, which is considered a "gentleman's trade." It is not surprising in the light of what we have said about the need for ocean-straddling trust in the TOMCs for auction crops.

11. The World Bank holds a different view. Around 1990, the World Bank urged African governments to have the local currency replaced by the dollar in the bidding at the auctions. This recommendation was implemented in several countries.

12. There are indications that, even before World War I, some importers of arabica coffee had become dissatisfied with the grades. Some British importers had employees in Brazil to buy coffee locally. The nascent coffee market in Kenya in the 1920s may have been modeled on the Brazilian example.

13. Selling locally may have been a special benefit for the Indian ginners in Uganda. The willingness of the European coffee growers in Kenya to join a cooperative union, the KPCU, is perhaps also indicative. Was it a recognition that most of them had little or no international reputation?

14. These timetables are particularly strict in the case of tobacco. In general, orderly marketing slows down the rate of evacuation and thus increases the aggregate amount of preshipment finance that is needed.

Chapter 13

1. Many crops discussed in earlier chapters are perishable at the moment of harvesting, but they are preserved by on-farm processing before they enter the TOMC.

2. Some of the pineapples in the European supermarkets have also come by air; for instance, some come from Kenya and Côte d'Ivoire. It is, however, exceptional for bananas to be carried by plane.

3. This is the first comparative study. Language problems have probably deterred other researchers.

4. The Belgians embarked on a banana project in the Belgian Congo, but it was so late that it had hardly developed before World War II disrupted it. Portugal had no need for bananas from Africa, as the fruit was grown on Madeira.

5. It is my impression that in the colonial period, the growers took the initiative. It was a case of forward integration in the TOMC. In the 1970s, the initiative generally came from the TNCs; this was backward integration.

Chapter 14

1. The following questions require further research. Does the institutional world market create high marketability? Or is there a third factor such as high gradability or high global turnover which explains both the institutionalization of the market and the high marketability?

2. The fact that the selection may be distorted by monopolistic elements is not further discussed here.

3. See Ingham (1995) for a recent summary on Prebisch, Singer, and Lewis. Lewis' views can be found in many of his publications. An early view appears in an unpublished paper of 1944.

4. Some time ago, the term "vertical export diversification" was introduced which necessitated the use of the adjective "horizontal" for what until then had been simply known as export diversification. Vertical export diversification is briefly referred to in Chapter 10.

5. See also Chapter 8 about withdrawal in the cotton sector.

6. This may happen when product A is produced by previously underutilized resources (land and/or labor), which is an element of the vent-for-surplus theory, as elaborated by Myint (1958).

7. As the criterion "ease of marketing" is fluid, the borderline between easy and difficult products is elusive.

8. The established exporters may try to bar new rivals by legal and other means.

9. Prevention of erosion seems to me a better (and more permanent) policy objective than export diversification; it requires regular monitoring of the export basket.

10. In French Cameroon, arabica coffee was introduced in the inter-war period, but it was not marketed through auctions.

11. There is an urgent need for more research on such farmers and planters in Atlantic Africa. In the same way, the sophisticated African farmers in Côte d'Ivoire and Kenya after independence deserve more attention for their effect on the export basket.

12. This term was also used for the economics of Uganda and Nyasaland/ Malawi in Indian Ocean Africa.

13. Presumably, it was this narrow commodity orientation which prompted Amin (1971) to speak of the blockaded or stunted development of West Africa.

14. My observations in Sierra Leone are the basis for this impression (see Chapter 11). It should be noted, however, that the general attitude in Africa was to trust the EMBs because they were national institutions and to distrust the TNCs because they were foreign. However, in regard to diversification, this trust turns out to have been misplaced. More research on the EMBs' attitude toward the minor crops is desirable.

15. It is desirable that countries whose baskets did not experience erosion in the 1960s and 1970s retrospectively study why this did not happen.

Chapter 15

1. There are a few international systems which collect data on the product flows in the middle section. There is first the stamp system of the International Coffee Organization designed to provide a foolproof method to match export and import data. When the system operated, the central administration in London was in a position to calculate the amount of coffee afloat on the ocean at any time. Second, the statistical office of the European Union has a system for matching the

export data of the ACP (African, Caribbean, and Pacific) countries with the import data of its member states retrospectively. This system is part of the Lomé Conventions and covers a wide range of agricultural and mineral products.

2. Existing theories recognize that the ideal outcome may be impaired by protective import tariffs and monopolistic tendencies, but the effect of unpredictable marketing costs has not yet been noticed.

3. The new international order may be said to have begun in 1866, when the first trans-oceanic cable was laid (to North America). For Africa, the new order began in 1880.

4. Perhaps the human factor is recognized as more important in institutional change. The new order was "shaped" by human minds and did not "evolve" as Lewis suggests.

5. For a recent survey of the literature, see Cooper (1993).

6. This is the definition of Kohls and Uhl (1990, p. 193) with the term "outlay" substituted for "dollar."

Bibliography

Abbott, J.C. ed. (1993). *Agricultural and Food Marketing in Developing Countries: Selected Readings*. Wallingford: C.A.B. International.

Abbott, J.C. and H.C. Creupelandt. (1966). *Agricultural Marketing Boards: Their Establishment and Operation*. Rome: FAO.

Acland, J.D. (1971). *East African Crops*. London: Longman.

Albion, R.G. (1959). *Seaports South of Sahara: The Achievements of an American Steamship Service*. New York: Appleton-Century-Crofts.

Amin, S. (1971). *L'Afrique de l'Ouest bloquée: L'économie politique de la colonisation 1880-1970*. Paris: Editions de Minuit.

Amin, S. (1972). "Underdevelopment and dependence in black Africa—Origins and contemporary forms," *Journal of Modern African Studies,* X (4): pp. 503-524.

Arhin, K. (1985). "The Ghana Cocoa Marketing Board and the farmer." In *Marketing Boards in Tropical Africa*, edited by K. Arhin, P. Hesp, and L. van der Laan. London: Kegan Paul, pp. 37-52.

Arhin, K., P. Hesp, and L. van der Laan, eds. (1985). *Marketing Boards in Tropical Africa*. London: Kegan Paul.

Assoumou, J. (1977). *L'Economie du cacao: Agriculture d'exportation et bataille du développement en Afrique tropicale*. Paris: Delarge.

Austen, R.A. (1987). *African Economic History: Internal Development and External Dependency*. London: Currey.

Barratt Brown, M. and P. Tiffen. (1992). *Short Changed: Africa and World Trade*. London: Pluto.

Bates, R.H. (1981). *Markets and States in Tropical Africa: The Political Basis of Agricultural Policies*. Berkeley: University of California Press.

Bauer, P.T. (1954). *West African Trade: A Study of Competion, Oligopoly and Monopoly in a Changing Economy*. Cambridge, England: Cambridge University Press.

Beckman, B. (1976). *Organising the Farmers: Cocoa Politics and National Development in Ghana*. Uppsala, Sweden: Scandinavian Institute of African Studies.

Béroud, F. (1994). "Réflexions sur l'organisation des filières d'Afrique francophone," *Marchés Tropicaux,* (MMDXLII): pp. 1585-1586.

Bessem, F. "The SODECOTON, the village associations, and rural development in Northern Cameroon." Forthcoming report.

Brett, E.A. (1973). *Colonialism and Underdevelopment in East Africa: The Politics of Economic Change 1919-1939*. London: Heinemann.

Bright, C. (1898). *Submarine Telegraphs: Their History, Construction and Working*. London: Crosby Lockwood.

Brown, I., ed. (1989). *The Economies of Africa and Asia in the Inter-War Depression*. London: Routledge.

Buell, R.L. (1928). *The Native Problem in Africa, Volume I and II*. Second Impression, 1965, London: Cass.

Cabot, J. (1965). *Le Bassin du Moyen Logone*. Paris: ORSTOM.

Cain, J.P. and A.G. Hopkins. (1993a). *British Imperialism: Innovation and Expansion 1688-1914*. London: Longman.

Cain, J.P. and A.G. Hopkins. (1993b). *British Imperialism: Crisis and Deconstruction 1914-1990*. London: Longman.

Carlsson, J. (1983). "The marketing of cash crops—the multinational versus the parastatal solution." Paper presented at a seminar on *Marketing Boards in Tropical Africa*, Leiden, Netherlands: African Studies Centre.

Cassady, R. (1967). *Auctions and Auctioneering*. Berkeley, CA: University of California Press.

Chamberlin, E.H. (1933). *The Theory of Monopolistic Competition*. Cambridge, MA: Harvard University Press.

Claessens, S. and R.C. Duncan, eds. (1993). *Managing Commodity Price Risk in Developing Countries*. Baltimore, MD: Johns Hopkins University Press.

Claessens, S. and P. Varangis. (1993). "Implementing risk management strategies in Costa Rica's coffee sector." In *Managing Commodity Price Risk in Developing Countries*, edited by S. Claessens and R.C. Duncan. Baltimore: Johns Hopkins University Press, pp. 185-205.

Clairmonte, F.F. (1975). "Bananas." In *Commodity Trade of the Third World*, edited by C. Payer. London: Macmillan, pp. 129-153.

Clairmonte, F. and J. Cavanagh. (1988). *Merchants of Drink: Transnational Control of World Beverages*. Penang, Malaysia: Third World Network.

Clements, F. and E. Harben. (1962). *Leaf of Gold: The Story of Rhodesian Tobacco*. London: Methuen.

"Cocoa man," *West Africa*, XLV, 1961, p. 653.

Commander, S., ed. (1989). *Structural Adjustment and Agriculture: Theory and Practice in Africa and Latin America*. London: Overseas Development Institute.

Cooper, F. (1993). "Africa and the world economy." In *Confronting Historical Paradigms: Peasants, Labor, and the Capitalist World System in Africa and Latin America*, edited by F. Cooper, A. F. Isaacman, F. E. Mallon, W. Roseberry, and S. J. Stern. Madison: University of Wisconsin Press, pp. 84-201.

Coquery-Vidrovitch, C. (1972). *Le Congo au temps des grandes compagnies concessionnaires 1898-1930*. Paris/The Hague: Mouton.

Coquery-Vidrovitch, C. (1975). "L'Impact des intérêts coloniaux: SCOA et CFAO dans l'Ouest Africain, 1910-1965," *Journal of African History*, XVI (4): pp. 595-621.

Cowen, M. and N. Westcott. (1986). "British imperial economic policy during the war." In *Africa and the Second World War*, edited by D. Killingray and R. Rathbone. New York: St. Martin's, pp. 20-67.

Dand, R. (1993). *The International Cocoa Trade*. Cambridge: Woodhead.

Davies, P.N. (1973). *The Trade Makers: Elder Dempster in West Africa, 1852-1972.* London: Allen and Unwin.

Davies, P.N. (1990). *Fyffes and the Bananas: Musa Sapientum: A Centenary History (1888-1990).* London: Athlone.

Daviron, B. and W. Fousse. (1993). *La compétitivité des cafés Africaines.* Paris: Ministère de la Coopération.

Delaporte, G. (1976). "La caisse de stabilisation et de soutien des prix des productions agricoles: Vingt années au service du planteur et de l'etat," *Marchés Tropicaux* (MDLXXXVII), pp. 959-978.

Deutsch, J.G. (1990). "Educating the middlemen: A political and economic history of statutory cocoa marketing in Nigeria, 1936-1947." PhD thesis, University of London.

Dijkstra, T. (1990). *Marketing Policies and Economic Interests in the Cotton Sector of Kenya.* Leiden, Netherlands: African Studies Centre.

Dijkstra, T. and H.L. van der Laan. (1990). "The future of Africa's raw material marketing boards: Will local factories make some of them redundant?" *Journal of International Food & Agribusiness Marketing,* II (2): pp. 47-75.

Drummond, I.M. (1972). *British Economic Policy and the Empire.* London: Allen and Unwin.

Drummond, I.M. (1974). *Imperial Economic Policy, 1917-1939.* London: Allen and Unwin.

Durand, H. (1957). *Essai sur la Conjoncture de l'Afrique Noire.* Paris: Dalloz.

Ehrlich, C. (1965). "The Uganda economy 1903-1945." In *History of East Africa, Vol. II,* edited by V. Harlow and E. M. Chilver. Oxford, England: Clarendon Press, pp. 394-475.

Ehrler, F. (1977). *Handelskonflikte zwischen Europaeischen Firmen und Einheimischen Produzenten in Britisch-Westafrika: Die 'Cocoa Hold-Ups' in der Zwischenkriegszeit.* Zurich, Switzerland: Atlantis.

Ehrlich, C. (1973). "Building and caretaking: Economic policy in British tropical Africa, 1890-1960," *Economic History Review,* Second Series, XXVI (4): pp. 649-667.

Ellis, F. (1980). *A Preliminary Analysis of the Decline in Tanzanian Cashewnut Production 1974-1979.* Dar es Salaam, Tanzania: Economic Research Bureau.

Ellis, S., ed. (1995). *Africa Now: People, Policies, and Institutions.* London: Currey.

Fieldhouse, D.K. (1971). "The economic exploitation of Africa: Some British and French comparisons." In *France and Britain in Africa,* edited by P. Gifford and W. R. Louis. New Haven, CT: Yale University Press, pp. 593-662.

Fieldhouse, D.K. (1994). *Merchant Capital and Economic Decolonization: The United Africa Company, 1929-1987.* Oxford, England: Clarendon.

Ford, V.C.R. (1955). *The Trade of Lake Victoria.* Kampala, Uganda: EAISR.

Forrest, D. (1985). *The World Tea Trade: A Survey of the Production, Distribution, and Consumption of Tea.* Cambridge, England: Woodhead-Faulkner.

Fry, R. (1976). *Bankers in West Africa: The Story of the Bank of British West Africa Limited.* London: Hutchinson Benham.

Gaitskell, A. (1959). *Gezira: A Story of Development in the Sudan.* London: Faber and Faber.

Galbraith, J.K. (1957). *American Capitalism: The Concept of Countervailing Power.* London: Hamish Hamilton.

Gereffi, G. and M. Korzeniewicz, eds. (1994). *Commodity Chains and Global Capitalism.* Westport, CT: Praeger.

Goddard, T.N. (1925). *The Handbook of Sierra Leone.* London: Grant Richards.

Gordon-Ashworth, F. (1984). *International Commodity Control: A Contemporary History and Appraisal.* London: Croom Helm.

Great Britain, Colonial Office (1938). *Report of the Commission on the Marketing of West African Cocoa.* London: HMSO. Usually called the Nowell Report.

Grosh, B. (1993). "Contract farming in Africa: An application of the new institutional economics," *Journal of African Economies,* III, (2): pp. 231-261.

Hailey, Lord (1957). *An African Survey: Revised 1956.* London: Oxford University Press.

Hancock, W.K. (1942). *Survey of British Commonwealth Affairs, Vol. II, Problems of Economic Policy, 1918-1939, Part 2.* London: Oxford University Press.

Hancock, W.K. (1943). *Empire in the Changing World.* Washington, DC: Penguin.

Hanisch, R. (1975). "Confrontation between primary commodity producers and consumers: The cocoa hold-up of 1964-65," *Journal of Commonwealth & Comparative Politics,* XIII (3): 242-260.

Hart, K. (1982). *The Political Economy of West African Agriculture.* Cambridge: Cambridge University Press.

Haviland, W.E. (1954). "The economic development of the tobacco industry of Northern Rhodesia." *South African Journal of Economics,* XXII (4): pp. 375-384.

Hazlewood, A. (1961). *The Economy of Africa.* London: Oxford University Press.

Helleiner, G.K. (1966). *Peasant Agriculture, Government, and Economic Growth in Nigeria.* Homewood, IL: Irwin.

Heyer, J. (1976). "The marketing system." In *Agricultural Development in Kenya: An Economic Assessment,* edited by J. Heyer, J.K. Maitha and W.M. Senga. Nairobi, Kenya: Oxford University Press, pp. 313-363.

Hill, M.F. (1949). *Permanent Way: The Story of the Kenya and Uganda Railway.* Nairobi, Kenya: East African Railways and Harbours.

Hill, M.F. (1956). *Planters' Progress: The Story of Coffee in Kenya.* Nairobi: Coffee Board of Kenya.

Hinds, A.E. (1986). "Colonial policy and the processing of groundnuts: The case of Georges Calil," *International Journal of African Historical Studies,* XIX, (2): pp. 261-273.

Hogendorn, J.S. (1978). *Nigerian Groundnut Exports: Origins and Early Development.* Zaria, Nigeria: Ahmadu Bello University Press.

Hopkins, A.G. (1973). *An Economic History of West Africa.* London: Longman.

Hopkins, A.G. (1976). "Imperial business in Africa. Part II: Interpretations," *Journal of African History,* XVII, (2): pp. 267-290.

Hopkins, A.G. (1978). "Innovation in a colonial context: African origins of the Nigerian cocoa-farming industry, 1880-1920." In *The Imperial Impact: Stud-*

ies in the Economic History of Africa and India, edited by C. Dewey and A. G. Hopkins. London: Athlone, pp. 83-96.

Houtkamp, J.A. (1996). *Tropical Africa's Emergence as a Banana Supplier in the Interwar Period.* Aldershot, England: Avebury.

Houtkamp, J.A. and H. L. van der Laan. (1993). *Commodity Auctions in Tropical Africa: A Survey of the African Tea, Tobacco, and Coffee Auctions.* Leiden, Netherlands: African Studies Centre.

Ingham, B. (1995). *Economics and Development.* London: McGraw-Hill.

Isaacman, A. (1996). *Cotton is the Mother of Poverty: Peasants, Work, and Rural Struggle in Colonial Mozambique, 1938-1961.* London: Currey.

Isaacman, A. and R. Roberts, eds. (1995). *Cotton, Colonialism, and Social History in Sub-Saharan Africa.* London: Currey.

Jaffee, S. (1995). "Private sector response to market liberalization in Tanzania's cashew nut industry." In *Marketing Africa's High-Value Foods: Comparative Experiences of an Emergent Private Sector,* edited by S. Jaffee and J. Morton. Dubuque, Iowa: Kendall Hunt, pp. 153-198.

Jaffee, S. and J. Morton. (1994). *Africa's Agro-entrepreneurs: Private Sector Processing and Marketing of High-value Foods.* AFTES Working Paper, Washington, DC: World Bank.

Johnson, E.A.J. (1970). *The Organization of Space in Developing Countries.* Cambridge, MA: Harvard University Press.

Kohls, R.L. and J.N. Uhl. (1990). *Marketing of Agricultural Products.* New York: Macmillan.

Kotler, P. (1994. *Marketing Management: Analysis, Planning, Implementation, and Control Eighth Edition.* Englewood Cliffs, NJ: Prentice-Hall.

Kriesel, H. (1969). *Cocoa Marketing in Nigeria.* Ibadan, Nigeria: NISER.

Latham, A. J. H. (1978). *The International Economy and the Undeveloped World 1965-1914.* London: Croom Helm.

Lele, U., N. Van de Walle, and M. Gbetibouo. (1989). *Cotton in Africa: An Analysis of Differences in Performance.* MADIA Discussion Paper, Washington, DC: World Bank.

Lesage, C. (1915). *Les Cables Sous-marins Allemands.* Paris: Plon-Nourrit.

Leubuscher, C. (1939). "Marketing schemes for native-grown produce in African territories." *Africa,* XII: pp. 163-187.

Leubuscher, C. (1951). *The Processing of Colonial Raw Material: A Study in Location.* London: HMSO.

Leubuscher, C. (1956). *Bulk Buying from the Colonies.* London: Oxford University Press.

Leubuscher, C. (1963). *The West African Shipping Trade, 1909-1959.* Leiden, Netherlands: Sythoff.

Lewis, W.A. (1978a). *Growth and Fluctuations 1870-1913.* London: Allen and Unwin.

Lewis, W.A. (1978b). *The Evolution of the International Economic Order.* Princeton, NJ: Princeton University Press.

Little, P.D. and M.J. Watts, eds. (1994). *Living under Contract: Contract Farming and Agrarian Transformation in Sub-Saharan Africa.* Madison, WI: University of Wisconsin Press.

Mackintosh, M. (1989). *Gender, Class and Rural Transformation: Agribusiness and the Food Crisis in Senegal.* London: Zed.

Mars, J. (1948). "Extra-Territorial enterprises." In *Mining, Commerce, and Finance in Nigeria,* edited by M. Perham. London: Faber and Faber, pp. 43-136.

Marshall, A. (1890). *Principles of Economics.* London: Macmillan.

Martin, A. (1963). *The Marketing of Minor Crops in Uganda.* London: HMSO.

Matheson, J.K. and E. W. Bovill, eds. (1950). *East African Agriculture.* London: Oxford University Press.

McCormack, R.L. (1976). "'Airlines and empires: Great Britain and the 'scramble for Africa'." *Canadian Journal of African Studies,* X (1): pp. 87-105.

McCracken, J. (1983). "'Planters, peasants, and the colonial state: The impact of the Native Tobacco Board in the Central Province of Malawi." *Journal of Southern African Studies,* 9: pp. 172-192.

Meredith, D. (1988). "'The Colonial Office, British business interests, and the reform of cocoa marketing in West Africa, 1937-1945," *Journal of African History,* 29: pp. 285-300.

Milburn, J. (1970). "'The 1938 Gold Coast cocoa crisis: British business and the Colonial Office," *African Historical Studies,* III (1): pp. 57-74.

Mitchell, B.R. (1982). *International Historical Statistics: Africa and Asia.* New York: New York University Press.

Mosley, P. (1983). *The Settler Economies: Studies in the Economic History of Kenya and Southern Rhodesia, 1900-1963.* Cambridge, England: Cambridge University Press.

Munro, J.F. (1976). *Africa and the International Economy since 1800.* London: Dent.

Munro, J.F. (1984). *Britain in Tropical Africa, 1880-1960.* London: Macmillan.

Myint, H. (1958). "'The 'classical theory' of international trade and the underdeveloped countries." *Economic Journal,* LXVIII: pp. 317-337.

Neumark, S.D. (1964). *Foreign Trade and Economic Development in Africa: A Historical Perspective.* Stanford, CA: Food Research Institute.

Newbury, C. (1978). "'Trade and technology in West Africa: The case of the Niger Company, 1900-1920," *Journal of African History,* XIX (4): pp. 551-575.

Nowell Report (1938). See Great Britain Colonial Office (1938).

Ohlin, B. (1933). *Interregional and International Trade.* Cambridge, MA: Harvard University Press.

Onitiri, H.M.A. and D. Olatunbosun, eds. (1974). *The Marketing Board System: Proceedings of an International Conference.* Ibadan, Nigeria: NISER.

Ord, H.W. and I. Livingstone. (1969). *An Introduction to West African Economics.* London: Heinemann.

Organization of African Unity. (1982). *The Lagos Plan of Action for the Economic Development of Africa 1980-2000.* Second (revised) edition. Geneva: International Institute for Labour Studies.

Palmer, R. and N. Parsons, eds. (1977). *The Roots of Rural Poverty in Central and Southern Africa*. Berkeley, CA: University of California Press.

Pedler, F. (1974). *The Lion and the Unicorn in Africa: A History of the Origins of the United Africa Company 1787-1931*. London: Heinemann.

Pedler, F. (1979). *Main Currents of West African History 1940-1978*. London: Macmillan.

Pedler, F. J. (1955). *Economic Geography of West Africa*. London: Longman.

Péhaut, Yves. (1973). *Les Oléagineux dans les pays d'Afrique occidentales associés au Marché commun: La production, le commerce et la transformation des produits*. Vols. I and II. Paris: Champion.

Phillips, A. (1989). *The Enigma of Colonialism: British Policy in West Africa*. London: Currey.

Pim, A. (1946). *Colonial Agricultural Production: The Contribution Made by Native Peasants and by Foreign Enterprise*. London: Oxford University Press.

Prebisch, R. (1962). "The Economic Development of Latin America and its Principal Problems." *Economic Bulletin for Latin America*. VII (1). The original publication appeared in 1950.

Ramboz, Y.-C. (1965). "La politique caféière de Côte d'Ivoire et la réforme de la caisse de stabilisation des prix du café et du cacao." *Revue Juridique et Politique, Indépendance et Coopération*, XIX, (2): pp. 194-218.

Ricardo, D. (1817). *The Principles of Political Economy and Taxation*. Reprinted 1973. London: Dent.

Robinson, J. (1933). *The Economics of Imperfect Competition*. London: Macmillan.

Sawadogo, A. (1977). *L'Agriculture en Côte d'Ivoire*. Paris: Presses Universitaires de France.

Scherer, F.M. and D. Ross. (1990). *Industrial Market Structure and Economic Performance*. Boston: Houghton Mifflin.

Scitovsky, T. (1952). *Welfare and Competition: The Economics of a Fully Employed Economy*. London: Allen and Unwin.

Shenton, R.W. (1986). *The Development of Capitalism in Northern Nigeria*. London: Currey.

Singer, H.W. (1950). "The distribution of gains between investing and borrowing countries." *American Economic Review*, XL (2): pp. 473-485.

Southall, R.J. (1978). "Farmers, traders, and brokers in the Gold Coast cocoa economy," *Canadian Journal of African Studies*, XII (2): pp. 185-211.

Stahl, K.M. (1951). *The Metropolitan Organization of British Colonial Trade*. London: Faber and Faber.

Stern, L.W. and A.I. El-Ansary. (1988, 1992). *Marketing Channels*. Third and Fourth Editions. Englewood Cliffs, NJ: Prentice Hall.

Stuerzinger, U. (1980). *Der Baumwollanbau im Tschad: Zur Problematik landwirtschaftlicher Exportproduktion in der Dritten Welt*. Zurich, Switzerland: Atlantis.

Suret-Canale, J. (1964). *Afrique noire, occidentale et centrale, Vol. II: L'ère coloniale, 1900-1945*. Paris: Editions sociales.

Swainson, N. (1980). *The Development of Corporate Capitalism in Kenya: 1918-1977*. London: Heinemann.

Terpstra, V. (1983). *International Marketing*. Chicago: Dryden.

Thomas, H.B. and R. Scott. (1935). *Uganda*. London: Oxford University Press.

Tosh, J. (1980). "The cash-crop revolution in tropical Africa: An agricultural reappraisal." *African Affairs*, LXXIX, (314), pp. 79-94.

Tousley, R.D., E. Clark, and F.E. Clark. (1962). *Principles of Marketing*. New York: Macmillan.

Tresselt, D. (1967). *The West African Shipping Range*. New York: United Nations.

UAC *Statistical and Economic Review*. London: United Africa Company, 1948-1963.

UNCTAD (1993). *Handbook of International Trade and Development Statistics 1993*. New York/Geneva: UNCTAD.

UNCTAD (1994). *National Institution Building to Facilitate Access to Risk Management Markets for Small Producers and Traders Particularly from Developing Countries and Countries in Transition: Issues Involved and Possible Ways to Overcome Them*. Geneva: UNCTAD.

UNECA (United Nations Economic Commission for Africa). (1989). *African Alternative Framework to Structural Adjustment Programmes for Socio-Economic Recovery and Transformation*. Addis Ababa, Ethiopia: UNECA.

Van Chi-Bonnardel, R. (1978). *Vie de relations au Sénégal: La Circulation Des Biens*. Dakar, Senegal: IFAN (Institut Fondamental d'Afrique Noire).

Van der Laan, H.L. (1965). *The Sierra Leone Diamonds*. London: Oxford University Press.

Van der Laan, H.L. (1975). *The Lebanese Traders in Sierra Leone*. Paris/The Hague: Mouton.

Van der Laan, H.L. (1983a). "Modern Inland Transport and the European Trading Firms in Colonial West Africa." *Cahiers d'Etudes Africaines*, 84: pp. 547-575.

Van der Laan, H.L. (1983b). "The European trading companies in West Africa: Their withdrawal from upcountry, 1945-1980." In *Entreprises et entrepreneurs en Afrique, XIXe et XXe siècles, Vol. II*. eds. Laboratoire 'Connaissance du Tiers-Monde.' Paris: Harmattan, pp. 369-384.

Van der Laan, H.L. (1984). "Trading in the Congo: The NAHV from 1918 to 1955." *African Economic History*, 12: pp. 241-259.

Van der Laan, H.L. (1986). "The selling policies of African export marketing boards." *African Affairs*, XXXV (340): pp. 365-383.

Van der Laan, H.L. (1987a). "Marketing West Africa's export crops: Modern boards and colonial trading companies." *Journal of Modern African Studies*, XXV (1): pp. 1-24.

Van der Laan, H.L. (1987b). "Selling tropical Africa's Export Crops: The experience of the interwar period." In *The State and the Market: Studies in the Economic and Social History of the Third World*, edited by C. Dewey. New Delhi, India: Manohar, pp. 238-261.

Van der Laan, H.L. (1988). *Cocoa and Coffee Buying in Cameroon: The Role of the Marketing Board in the South West and North West Provinces, 1978-1987.* Leiden, Netherlands: African Studies Centre.

Van der Laan, H.L. (1989). "Export crop marketing in tropical Africa: What role for private enterprise?" *Journal of International Food & Agribusiness Marketing,* I (1): pp. 41-62.

Van der Laan, H.L. (1991). "A macro-economic model of the colonial peasant economy, with special reference to West Africa." Paper presented at the Third World Economic History and Development Group Conference in Manchester.

Van der Laan, H.L. (1992). "In defence of export marketing boards," *APROMA Bimonthly Review,* (27/28/29): pp. 34-40.

Van der Laan, H.L. (1993a). "Boosting agricultural exports? A 'marketing channel' perspective on an African dilemma." *African Affairs,* XCII (367): pp. 173-201.

Van der Laan, H.L. (1993b). "Recommending World Market Prices as Signals to Small Farmers in Africa," *APROMA Bimonthly Review:* (33), 3-9.

Van der Laan, H.L. and W.T.M. Van Haaren. (1990). *African Marketing Boards under Structural Adjustment: The Experience of Sub-Saharan Africa during the 1980s.* Leiden, Netherlands: African Studies Centre.

Van Zwanenberg, R.M.A. (1975). *An Economic History of Kenya and Uganda, 1800–1970.* London: Macmillan.

Vloeberghs, H. (1956). "La stabilisation des prix des produits agricoles en Afrique." In *Vers la promotion de l'économie indigène,* edited by Colloque Colonial sur l'Economie Indigène. Brussels: Institut Sociologie Solvay, pp. 121-144.

Walker, G. (1959). *Traffic and Transport in Nigeria.* London: HMSO.

Westcott, N. (1984). "The East African sisal industry, 1929-1949: The marketing of a colonial commodity during depression and war." *Journal of African History,* XXV (4): pp. 445-461.

Westcott, N. (1987). "Stabilizing commodity prices: State control of colonial commodity trade, 1930-1950." In *The State and the Market: Studies in the Economic and Social History of the Third World,* edited by C. Dewey. New Delhi: Manohar, pp. 262-287.

Wickizer, V.D. (1951). *Coffee, Tea, and Cocoa: An Economic and Political Analysis.* Stanford, CA: Stanford University Press.

Williams, I.A. (1931). *The Firm of Cadbury, 1831-1931.* London: Constable.

Williams, J.C. and B.D. Wright. (1991). *Storage and Commodity Markets.* Cambridge, England: Cambridge University Press.

Williamson, O.E. (1975). *Markets and Hierarchies: Analysis and Antitrust Implications.* New York: Free Press.

World Bank. (1981). *Accelerated Development in Sub-Saharan Africa: An Agenda for Action.* Washington, DC: World Bank.

World Bank. (1984). *Towards Sustained Development in Sub-Saharan Africa.* Washington, DC: World Bank.

World Bank. (1986). *Financing Adjustment with Growth in Sub-Saharan Africa.* Washington, DC: World Bank.

World Bank (1989). *Sub-Saharan Africa: From Crisis to Sustainable Growth: A Long-term Perspective Study.* Washington, DC: World Bank.

World Bank. (1994). *Adjustment in Africa: Reforms, Results, and the Road Ahead.* Washington, DC: World Bank.

Wrigley, C.C. (1965). "Kenya: The patterns of economic life, 1902-1945." In *History of East Africa Vol. II,* edited by V. Harlow and E.M. Chilver. Oxford, England: Clarendon, pp. 209-264.

Wrigley, C.C. (1990). "Aspects of economic history." In *The Colonial Moment in Africa: Essays on the Movement of Minds and Materials 1900-1940,* edited by A. Roberts. Cambridge, England: Cambridge University Press, pp. 77-139.

Yoshida, M. (1984). *Agricultural Marketing Intervention in East Africa.* Tokyo: Institute of Developing Economies.

Index

Page numbers followed by the letter "f" indicate figures; those followed by the letter "t" indicate tables.

Order Your Own Copy of This Important Book for Your Personal Library!

THE TRANS-OCEANIC MARKETING CHANNEL
A New Tool for Understanding Tropical Africa's Export Agriculture

_____ in hardbound at $49.95 (ISBN: 0-7890-0116-0)

COST OF BOOKS_____

OUTSIDE USA/CANADA/
MEXICO: ADD 20%_____

POSTAGE & HANDLING_____
(US: $3.00 for first book & $1.25
for each additional book)
Outside US: $4.75 for first book
& $1.75 for each additional book)

SUBTOTAL_____

IN CANADA: ADD 7% GST_____

STATE TAX_____
(NY, OH & MN residents, please
add appropriate local sales tax)

FINAL TOTAL_____
(If paying in Canadian funds,
convert using the current
exchange rate. UNESCO
coupons welcome.)

☐ **BILL ME LATER:** ($5 service charge will be added)
(Bill-me option is good on US/Canada/Mexico orders only;
not good to jobbers, wholesalers, or subscription agencies.)

☐ Check here if billing address is different from
shipping address and attach purchase order and
billing address information.

Signature_____

☐ **PAYMENT ENCLOSED: $**_____

☐ **PLEASE CHARGE TO MY CREDIT CARD.**

☐ Visa ☐ MasterCard ☐ AmEx ☐ Discover
☐ Diners Club
Account # _____

Exp. Date _____

Signature _____

Prices in US dollars and subject to change without notice.

NAME _____

INSTITUTION _____

ADDRESS _____

CITY _____

STATE/ZIP _____

COUNTRY _____ COUNTY (NY residents only) _____

TEL _____ FAX _____

E-MAIL_____
May we use your e-mail address for confirmations and other types of information? ☐ Yes ☐ No

Order From Your Local Bookstore or Directly From
The Haworth Press, Inc.
10 Alice Street, Binghamton, New York 13904-1580 • USA
TELEPHONE: 1-800-HAWORTH (1-800-429-6784) / Outside US/Canada: (607) 722-5857
FAX: 1-800-895-0582 / Outside US/Canada: (607) 772-6362
E-mail: getinfo@haworth.com
PLEASE PHOTOCOPY THIS FORM FOR YOUR PERSONAL USE.

BOF96